Computers and Writing

The Cyborg Era

Computers and Writing

The Cyborg Era

James A. Inman
University of South Florida

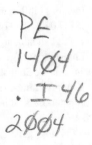
 LAWRENCE ERLBAUM ASSOCIATES, PUBLISHERS
2004   Mahwah, New Jersey   London

Lawrence Erlbaum Associates, Inc., Publishers
10 Industrial Avenue
Mahwah, NJ 07430

Cover design by Kathryn Houghtaling Lacey

Library of Congress Cataloging-in-Publication Data

Inman, James A.
   Computers and writing : the cyborg era / James A. Inman
     p.   cm.
   Includes bibliographical references and indexes.
   ISBN 0–8058–4160–1 (alk. paper) — ISBN 0–8058–4161–X (pbk. : alk. paper)
   1. English language—Rhetoric—Study and teaching—Data processing.
2. English language—Rhetoric—Study and teaching—Computer network
resources. 3. English language—Rhetoric—Computer-assisted instruction.
4. Report writing—Study and teaching—Data processing. 5. Report writing—
Computer-assisted instruction. 6. Report writing—Computer network resources.
I. Title.
PE1404 .I46 2003
808'.042'0285—dc21
                                        2002192536

Books published by Lawrence Erlbaum Associates are printed on acid-free paper,
and their bindings are chosen for strength and durability.

Printed in the United States of America
10 9 8 7 6 5 4 3 2 1

Contents

Foreword

Anne Ruggles Gere
University of Michigan, Ann Arbor

These days the word *computer* is frequently followed by *posthuman,* a term that has already taken on multiple meanings including human identity as information pattern rather than embodied action, human nature transformed by biotechnology, or the material human body as something to be redesigned or left behind altogether. Accordingly, it is refreshing to open a book that insists computers must be considered in light of the people who work with them, and the human community is every bit as important as the technology.

James Inman uses a variety of strategies to make good on his claims for the importance of the human community of computers and writing. I didn't count, but he includes many, many people from the community. Some of them respond to questions, some appear in a montage of scholarly voices, and some participate in an electronic conversation. The questions include "How did you come to be active in the computers and writing community?" "What is the best lesson you have learned from the computers and writing community?" "What scholarly project in computers and writing has been most influential for you, and why has it been so influential?" Some responses are highly personal accounts, whereas others offer a perspective on the computers and writing community, but there are no commentaries on the posthuman.

The question I found most interesting was "What worries you about the computers and writing community, and why does it worry you?" Responses varied from what I would call status concerns—how computers and writing does or does not connect with other communities, fields, and disciplines in the academy—to wondering where the community will go next and if it will be able to retain its cutting edge quality, to concerns about how little the community problematizes software, to questions about what technology does for students. No one worries about the implications of posthumanism.

The montage of scholarly voices includes a variety of pronouncements on what the computers and writing community is accomplishing, its varying levels of effectiveness, and issues it will need to face. Participants in the Tuesday Café discussion of the Netoric Project focus on the future of computers and writing, considering technical support, compensation, and disagreements within the community. There's nothing about the posthuman in either of these, or anywhere else in the book.

Given both Inman's argument for the importance of the human community and the way he makes that argument, the absence of the posthuman is not surprising. Instead of looking toward the ways human nature will be changed by technology, he does several things to underscore the community among the humans represented in the book. In addition to including comments by a great number of individuals, Inman presents everyone in a nonhierarchical way. None of the usual markers of institutional affiliation and status are included. Graduate students appear alongside graying professors, and the comments of each receive equal attention. What Inman does include is a photograph of each of the individuals who responds to a question in one of the "Community Voices" sections. We see faces and bodies along with the words.

In her book, *How We Became Posthuman,* N. Katherine Hayles (1999) observes that thought depends on the specificities of the embodied form enacting it. She argues that it does matter that emotions in humans are mediated through the hormonal system, whereas emotions in computers are created through feedback loops between algorithmically encoded goals, scripts, and personality parameters. Inman's inclusion of images of the embodied selves who participate in the computers and writing community speaks directly to that issue by displaying images of the embodied forms that generated the thought encoded on the page. Furthermore, Inman's concept of the *cyborg era,* which he defines as a present in which individuals, technologies, and their contexts all receive equal attention, insists that people have a necessary relationship to their technologies. Because the term *cyborg era* builds on the concept of the *cyborg as an agent of change,* the human–technology relationship in this view will include agency for the human.

With that agency comes an agenda which Inman details as (a) remembering individuals in any technology and/or technology-adoption decision, (b) actively seeking and promoting diversity, (c) articulating and modeling resistance, and (d) participating in the design of technologies. In many ways this book shares a common perspective with what is called *humanistic informatics,* a broad interdisciplinary area that studies the interactions between humans, the institutions they create, and the information technologies, especially computers. Regardless of what we call it, Inman's perspective offers an important alternative to those who elide what Hayles (1999) calls the very real and significant differences in embodiment between protein and silicon life forms.

Preface

Computers and Writing: The Cyborg Era explores the landscape of the contemporary computers and writing community. Its six chapters engage critical issues in the community, including redefining its generally accepted history, connecting contemporary innovators with the community's longstanding spirit of innovation, advocating for increased access and diversity through pedagogy, and more. Between chapters, readers will find "Community Voices" segments, which are designed to provide a snapshot of the contemporary computers and writing community and to introduce community members in their own voices. These elements together define what I am calling *the cyborg era* of computers and writing.

The defining features of *Computers and Writing* are its introduction of the cyborg era as a term requiring consistent and equitable attention on individuals, technologies, and the contexts they share all at once, rather than on only one or two such elements, and its argument that the computers and writing community is best represented by many voices, rather than a single perspective or series of perspectives. These emphases inform and indeed shape each of the six chapters, as well as the "Community Voices" segments. No previous project has presented the issues and individuals of the community all at once and at such a broad scale, as this book does. In this way, *Computers and Writing* is a unique venture.

As already noted, *Computers and Writing*'s chapters address key issues in the contemporary computers and writing community:

- Chapter 1 defines *computers and writing* as a community, emphasizing the individuals who interact and do work in the area, rather than the work they do alone. The chapter also defines the cyborg era, arguing that it reflects a necessary and compelling contemporary convergence of individuals, technologies, and the contexts they share. Chapter 1 concludes with a montage of scholarly views, emphasizing the many voices of the community in form and content.

- In chapter 2, I argue for an extended history of the computers and writing community, one grounded not in when computers first entered writing classrooms, but when individuals first began interacting with each other and with technology in all of its many forms. I emphasize the 1960s and 1970s as especially telling decades, arguing we should consider them foundational in the community's history.

• Chapter 3 conceptualizes canonized computers and writing scholarship as cyborg narratives, threads of knowledge about community innovations that can provide encouragement, support, and context for those doing too often unappreciated or underrespected innovative work in the contemporary era.

• In chapter 4, I present a new term, *cyborg literacy,* to represent contemporary meaning-making practices, placing it in conversation with existing terms like *electronic literacies, computer literacy,* and *technology literacy* and then developing it more fully by presenting research about the advantages its broad definition offers.

• Chapter 5 offers *cyborg pedagogy* as an activist teaching approach, one designed to address the access and diversity issues that any contemporary use of technology necessarily brings. To bring cyborg pedagogy as a term into specific relief, I present analysis of a representative assignment that I introduced into a class at Furman University.

• Chapter 6 outlines a specific agenda for the future of the computers and writing community by introducing the notion of *cyborg responsibility,* which is essentially responsibility to question accepted beliefs and practices and to serve as an activist in and beyond the community. It includes excerpts of a MOO conversation about the future of the community, again reflecting the book's emphasis on many voices, and it articulates specific agenda elements, seeking to provide readers with ideas they can pursue.

These chapters together provide readers with a sense of the most compelling issues in the contemporary computers and writing community, and it is my hope that the chapters create new conversations and/or energize existing conversations about these key issues.

The "Community Voices" segments may be between chapters, but they are vitally important to the book because they offer readers the chance to meet and hear from many individuals in the computers and writing community around the world. In addition to providing pictures, the individuals responded to one of six questions that I posed:

• How did you come to be active in the computers and writing community?
• What scholarly project in computers and writing has been most influential for you, and why has it been so influential?
• What's the most important aspect of the computers and writing community for you, and why is it so important?
• What worries you about the computers and writing community, and why does it worry you?

- What's the best lesson you've learned from the computers and writing community, and why is it the best?
- Why do you choose to be active in the computers and writing community?

The representation in the "Community Voices" segments is truly global; community members whose voices are included represent six different continents in sum. The segments also present the dynamic, ever-evolving nature of the computers and writing community; some of the individuals have been extremely active in recent community-specific events, whereas others have not, and I believe it's important to include them all as a way of best representing the community's many voices. Everyone doing important and influential work will not attend every related conference or publish in every related journal, but their work is vital nonetheless.

ACKNOWLEDGMENTS

This project has benefited from the guidance, encouragement, and support of many people, and I'd like to acknowledge them. Although I name specific individuals here, I mean this also to be a general acknowledgment of everyone who has helped along the way, whether specifically named or not. You have made an amazing difference for this book and for me.

A book about the computers and writing community requires active support from community members, and indeed many outstanding colleagues and friends came through for me, taking time from their already too busy schedules to author responses for the "Community Voices" sections that readers will find between chapters. These individuals trusted me not only with their words, but also with pictures of themselves, and I'm very cognizant of the degree of trust that reflects and extremely grateful to all of them. Specific credits for photos should go to Ruth Ellen Kocher, Bruce Leland, and Helga Mayringer, in addition to others whose names I do not know.

Throughout the development of the book, I was fortunate to work with a number of colleagues and friends, who read drafts and offered insightful feedback. I should begin with my dissertation committee at the University of Michigan, as the book began there: Anne Ruggles Gere, Barbara Monroe, Lesley Rex, Marlon Ross, and Cindy Selfe. I am especially excited that Anne has written a foreword for the book, as she has meant so much to this project and to me; I cannot imagine a more amazing mentor. A number of colleagues—including Anne and Cindy—very generously read individual chapter drafts, as I continued to work on the project, including Lloyd Benson, Joanna Castner, Dagmar Corrigan, Lisa Gerrard, Simone Gers, Glen Halva-Neubauer, Gail Hawisher, Krista Homicz, Lisa Lebduska, Lynley Loftin, Julia Makosky, Leigh Ryan, Pete Sands, Donna Sewell, Lynne Shackelford, and Stone Shiflet. The outside readers commissioned by Lawrence Erlbaum Associates, Hugh Burns (Texas Women's University), Barry Maid (Arizona State University–East), and Carl Whithaus (Old Dominion University), also provided valuable feedback. I'm grateful for all that everyone has done; this project would not have been the same with you.

I would also like to thank my family, as they've been a constant and encouraging presence throughout the duration of this project and indeed my entire life: my parents Ralph and Sandra Inman, grandparents Ralph and Ellen Inman and Mack and Anne Wood, and uncle and aunt Jim and Pam Inman.

Last, but certainly not least, I'd like to thank the editorial and production staff at Lawrence Erlbaum Associates. I've worked with Naomi Silverman on three book projects now, and I've found her to be exceptional each time, with assistant editor Lori Hawver and editorial assistants Stacey Mulligan and Erica Kica. The production team for this book has also been fantastic.

—James A. Inman

1

Defining Computers and Writing: Defining the Cyborg Era

Computers and Writing: The Cyborg Era is a book that explores the landscape of the contemporary computers and writing community, emphasizing the challenges and opportunities before us. The book's chapters are organized into three sections—past, present, and future—and each chapter presents a new approach to thinking about theory and practice in computers and writing, from questioning and extending what we know of its history, to shaping instances of its pedagogy to promote equity and diversity. I write as a practitioner, a participant–observer with a professional and personal investment in the computers and writing community's growth and development.

In this chapter, I explore both of *Computers and Writing*'s key terms: *computers and writing* and *the cyborg era*. I begin by defining computers and writing as a community, suggesting that it should be represented by its members and their interactions and resources. With that basis, I then argue that contemporary computers and writing is in a "cyborg era," a period name that emphasizes individuals, technologies, and the contexts they share simultaneously, rather than just technologies or individuals. Last I present a montage of excerpts from scholarly projects in the cyborg era of computers and writing, inviting readers to draw connections with me across the perspectives and opinions woven together.

DEFINING COMPUTERS AND WRITING

Thinking about computers and writing is certainly not new. In 1983, Kathleen Kiefer and Cynthia L. Selfe founded the print journal *Computers and Composi-*

1

tion, a publication that remains prominent today. Although it would be a mistake to conflate the terms *composition* and *writing,* they do prove close in application; many articles in *Computers and Composition* have identified writing theory and practice as their emphasis, rather than composition theory and practice. Also in 1983, an annual computers and writing conference tradition was born, when Lillian Bridwell-Bowles welcomed scholars to Minneapolis, Minnesota. With the exception of 1987 and 1988, the conference has been held every year since, and it has grown substantially both in size and scope. Computers and writing also gained attention with the publication of two books in the early 1990s: *Computers and Writing: Theory, Research, Practice,* an edited collection by Deborah Holdstein and Cynthia L. Selfe (1990), and *Computers and Writing: State of the Art,* an edited collection by Patrik Holt and Noel Williams (1992). Additionally, I would point to *Computers and the Teaching of Writing in American Higher Education, 1979–1994: A History* by Gail E. Hawisher, Paul LeBlanc, Charles Moran, and Cynthia L. Selfe, published in 1996. The authors' intent was not to chronicle the history of computers and writing, but the book has often been read and cited that way. In a review published in *Kairos: A Journal for Teachers of Writing in Webbed Environments,* for instance, Kip Strasma (1997) writes that *History* should be "required reading" for graduate courses ". . . not because it is the first attempt at a complete history, but because it has completed its history so well. Although not comprehensive, as the writers note that no history can be, it presents a thorough account of the key events, publications, and professionals who are responsible for making the field what it is today." One of the interesting elements of *History* is how it challenges readers to learn more; Lisa Gerrard's (1996) preface identifies K–12 educators' innovations with technology in the 1960s and 1970s as also important to know, and the book's focus on American higher education clearly does not address key advancements outside of the United States.

Despite the scholarly activity and attention, we still have uncertainty and even conflict about what computers and writing means. Some scholars have termed it a field or discipline (Condon, 2000; Lang, 2000; LeBlanc, 1998; Rodrigues, 2001; Slatin, 1998), whereas others have referred to it as a subfield or subdiscipline of a more general area of study (Haas, 1996; Hawisher et al., 1996; Schwalm, 1998); still more have suggested that computers and writing is a community (Gerrard, 1998; Rosenberg, 1999). In pointing out the conflict, I do not mean to imply that one and only one definition should be accepted; clearly any single definition would exclude as many people and projects as it would include. We have to realize, however, that terms like *field, discipline, subfield, subdiscipline,* and *community* are not interchangeable, as they each bring forward distinct values and implications. Terming *computers and writing* a field, for instance, suggests that it has an established unique body of scholarship and that a number of scholars are engaged in its work, developing new scholarship themselves that advances knowledge in the field. On the other hand, representing computers and writing as a community means that its character is determined by the relationships and resources individuals share. Understanding differences like these affords us a richer understanding of how computers and writing has been and is still being defined, thus enabling us to develop our individual perspectives in more careful, informed ways. Diversity of views is important, as is creating contexts in which multiple ones may be considered respectfully, but we have a responsibility to be certain that whatever definition or definitions we adopt are developed thoroughly and with attention to their critical implications.

Of the definitions scholars have suggested, I believe computers and writing is best represented as a community because that definition foregrounds the individuals involved in its work, rather than the scholarship or other knowledge they present. Even if I wasn't determined to emphasize the individuals in computers and writing, though, I do not feel comfortable thinking about computers and writing scholarship as distinct, separable from that of related areas, like rhetoric and composition, technical communication, information studies, and communication studies. And I believe that attempting to locate computers and writing as a subfield or subdiscipline of one such area is overly restrictive, not adequately representing its scope or depth. In defining computers and writing as a community, I point to three specific aspects of its activity:

- Real and virtual conferences;
- Professional organizations and initiatives; and
- Publishing ventures and products.

Each of these areas reflects an emphasis on individuals and their relationships and resources.

Real and Virtual Conferences

Computers and writing community events provide opportunities for members to interact with each other and for new members to be welcomed into the community. The most obvious place to begin is with the annual computers and writing conference, held in the spring of each year. The following table outlines the conference's history:

Year	Location	Coordinators
2002	Normal, Illinois	Ron Fortune and James Kalmbach
2001	Muncie, Indiana	Linda Hanson and Rich Rice
2000	Fort Worth, Texas	Dene Grigar, John F. Barber, and Hugh Burns
1999	Rapid City, South Dakota	Michael Day
1998	Gainesville, Florida	Anthony Rue
1997	Honolulu, Hawaii	Judith Kirkpatrick
1996	Logan, Utah	Christine Hult

(continued)

Year	Location	Coordinators
1995	El Paso, Texas	Evelyn Posey
1994	Columbia, Missouri	Eric Crump
1993	Ann Arbor, Michigan	Bill Condon
1992	Indianapolis, Indiana	Helen Schwartz, Linda Hanson, and Webster Newbold
1991	Biloxi, Mississippi	Rae Schipke
1990	Austin, Texas	Fred Kemp, John Slatin, Wayne Butler, and Locke Carter
1989	Minneapolis, Minnesota	Geoff Sirc, Trent Batson
1987–1988	No Conferences Held	
1986	Pittsburgh, Pennsylvania	Glynda Hull
1985	Los Angeles, California	Lisa Gerrard
1984	Minneapolis, Minnesota	Donald Ross
1983	Minneapolis, Minnesota	Lillian Bridwell-Bowles

Because the conference is so late in the year, it speaks especially well for the strength and vibrancy of computers and writing as a community. Few individuals can obtain travel funding, especially if they've attended other conferences throughout the year, so their attending the computers and writing conference reflects not just their engagement with the community's areas of study, but also their personal financial commitment to the value of the conference. Correspondingly, conference organizers have attempted to create cost-effective conference experiences. For instance, the 1998 conference was hosted by the University of Florida, and conference organizer Anthony Rue and his team provided attendees the option of staying in dormitories rather than at the conference hotel.

One of the most important elements of the computers and writing conference is the way it facilitates networking for individuals with mutual interests. Many in the computers and writing community work on the margins of their institutions, finding little help and support in institutional contexts, so interacting with others facing similar challenges makes a real difference. At the computers and writing conference, the achievements and perspectives of such individuals—often adjuncts and other part-time faculty members, staff, and graduate students—typically draw considerable attention, the sort of attention their work should be given in all contexts, but realistically isn't. One of the most prominent elements of recent computers and writing conferences has been the "town hall" meetings, sessions where all attendees gather in an auditorium and an extended panel of colleagues presents short statements about issues in computers and writing, then engages in a conversation with the audience. The town hall concept was introduced by Dene Grigar, John F. Barber, and Rebecca Rickly at the 1998 conference, and it has been a component of every conference since. Compelling also is that the computers and writing conference has stayed relatively small. Each year, the conference seems to grow, but it has not yet reached a point where members of the community can't find and visit with specific colleagues.

Although the computers and writing conference has provided important opportunities for individuals in the community, it cannot effectively reach everyone interested in attending. Responding to this challenge, conference organizers have created online conference opportunities, opportunities that likewise reflect a community view of computers and writing. As early as 1993 and 1994, organizers like Bill Condon and Eric Crump provided electronic discussion opportunities to unite those who would be attending the conference and those who could not, and these proved valuable for all involved. At the same time, the computers and writing community clearly could do more, and conference organizers began to explore how to create unique opportunities for individuals unable to attend the onsite conference. In 1999, organizer Michael Day asked Tari Lin Fanderclai, Sharon Cogdill, and me to develop online events, and we created an entire online conference experience, including presentations published on the World Wide Web, synchronous discussions with authors of the presentations, and an electronic list featuring special guests in discussions about various topics. Fanderclai and I teamed with Bradley Dilger and Cynthia A. Wambeam to offer further-expanded versions of the online conference in 2000 and 2001, bringing more and more individuals into the computers and writing community. In 2000, we were also able to create a formal selection process for hosting the computers and writing online conference to parallel the selection process for hosting the onsite conference, both processes being coordinated by the Conference on College Composition and Communication's Committee on Computers in Composition and Communication, or 7Cs. This formalized process brought about the distinction, "computers and writing onsite conference," which is the one that has been held since 1983, and "computers and writing online conference." The first online conference coordinator selected under this new system, Dene Grigar, put together a team that broadened the opportunities the conference provides for the computers and writing community by including video components and emphasizing archiving.

Although I begin with the annual onsite and online computers and writing conferences, I do not imply that they are the only conferences of note at which the computers and writing community is evident and exerts influence. On the contrary, community members have proven visible at a broad range of conference venues, and these venues have provided important opportunities for interaction between community members, as well as between community members and others interested in similar issues. At the 1998 convention of the Conference on College Composition and Communication (CCCC), for instance, Cynthia L. Selfe (1998) delivered the keynote address and emphasized the critical implications of computers in composition and communication. At the 2001 CCCC convention, Victor Vitanza (2001) offered a plenary address, "How Electronic Texts and Journals Will Shape Our Professional Work," in which he discussed the importance of electronic publishing. Additionally, the CCCC proposal form now includes a technology strand, ensuring that a significant number of panel presentations and workshops are present each year in the official program. Even more broadly, conventions of the National Council of Teachers of English (NCTE) and International Reading Association (IRA) have often featured computers and writing community members. John F. Barber and Dene Grigar have often joined colleagues like Michael Day, Joel English, Judi Kirkpatrick, and Ted Nellen to offer postconvention workshops on computers and writing at NCTE, and the associated Assembly for Computers in English (ACE) has coordinated a booth designed to offer NCTE attendees a chance to learn more about computers and writing. Hypertext scholars like Mark Bernstein, Jane Yellowlees Douglas, and Deena Larson have been active in conferences associated with the Association for Computing Machinery (ACM), especially the international annual Hypertext conference, taking on leadership roles and offering insightful presentations. Conferences for writing center scholars have also featured computers and writing community members, such as Donna N. Sewell, Michael Pemberton, and Clinton Gardner, as well as Barry M. Maid and Jennifer Jordan-Henley. The call for papers for the 2002 conference of the International Writing Centers Association (IWCA) even included two distinct technology strands: "technology for beginners" and "advanced technology." Last, online conferences have featured computers and writing community members. Two of special note have been the Teaching in the Community Colleges Online Conference (TCC Online), first held in 1996, and TeacherFest, held in 1999. Indeed at the first TCC Online conference, Eric Crump offered a keynote address, "Interversity: Convergence & Transformation, or Discovering the Revolution That Already Happened," and Judi Kirkpatrick coordinated MOO presentations.

Professional Organizations and Initiatives

Beyond conferences, other key factors emerge for thinking about computers and writing as a community. One of the first is that several organizations have been introduced. Perhaps the most widely known, the Alliance for Computers and Writing, or ACW, was founded in 1994 by Trent Batson and Fred Kemp as a way to bring together scholars interested in computers and writing. As described on the ACW's website (2001), it is

> . . . a national, non-profit organization committed to supporting teachers at all levels of instruction in their intelligent, theory-based use of computers in writing instruction. The operating principle behind ACW/Web is that writing teachers will provide the shared knowledge necessary for doing their job well if someone gives them the means to share that knowledge.

In truth, the influence of the ACW has waned in recent years, as it has become less and less active as an organization; however, it does represent one prominent beginning toward computers and writing as a community. It also remains the official sponsor of the journal, *Kairos*. Interestingly, some regional variations of the ACW have stayed active, outliving the larger organization in terms of influence. Two of special note are the Mid-Atlantic Alliance for Computers and Writing (MAACW) and the Great Plains Alliance for Computers and Writing (GPACW), both of which still host an annual conference to unite scholars in that region interested in computers and writing.

Another organization of note for the computers and writing community is the Netoric Project, founded in 1993 by Tari Lin Fanderclai and Greg Siering and now coordinated by Fanderclai, Siering, Cynthia A. Wambeam, and me. The organization's goal is ". . . to bring together geographically distant colleagues to discuss issues related to computer-assisted writing instruction," and its most popular event is the Tuesday Café, a weekly MOO discussion of computers and writing issues (2001). Netoric seeks especially to promote computers and writing as a community by welcoming new members into the fold and supporting their learning. One Tuesday Café per month is currently devoted to MOO skills, for instance, helping everyone in attendance learn more about MOO characteristics and conventions, and we've also run Netoric "How to MOO for the Tuesday Café" workshops at the past two computers and writing onsite conferences, a tradition we plan to continue at both onsite and online conferences in the future. Such efforts are critical, we believe, because they provide interested individuals with guided hands-on instruction in learning to MOO both in terms of technical requirements and operations and of conventions for interaction. Recent Café topics have included "MP3s and the Computers and Writing Professional," "Creative Writing in Electronic Spaces," "Technical Support: What Computers and Writing Faculty Should and Should Not Do," and "Adjunct and Other Part-Time Faculty in Computers and Writing: What We Can Do to Make a Difference."

The ACW and the Netoric Project both seek to be international in scope, and they have found success at times. It would be a mistake, however, to assume that they offer international computers and writing community members all that they need. In actuality, a number of organizations have been created in specific international contexts, and they have often flourished. For years, scholars in Australia have been doing prominent work in literacy studies, and a number like Ilana Snyder and Wendy Morgan have focused at least part of their professional efforts on the influence of computers and other technologies. Organizations such as the Australian Literacy Educators' Association (ALEA) and Australian Association for the Teaching of English (AATE) provide important opportunities for interaction among members with mutual interests; the organizations are large enough also to enable members interested in computers and other technologies to find new colleagues across Australia and even around the world. ALEA and AATE have conducted several joint conferences, including the 2002 event themed "e-volving literacies," which specifically emphasized technologies. Another example of international organizations is the Writing and Computers Association, founded in 1991 at the "Computers and Writing IV" conference at the University of Sussex in Great Britain, the conference itself having first been held in 1988. Though it remains a relatively small organization, its goals are significant, including to "promote worldwide the study of writing and design of computer-supported systems for writing." Surprisingly, there seems to have been very little interaction among the Writing and Computers Association and what I am calling the computers and writing community in official terms, despite their common goals and interests and their nearly identical names; scholars in the United States may not be able to name important scholars in Britain, that is, and vice versa. I read all of the associations named in this paragraph into this chapter's definition of computers and writing, however, because I believe we must begin to interact regularly, thinking about the community as global, including the organizations named as well as others around the world.

Other types of organizations that have become especially prominent for the computers and writing community are electronic lists, or e-lists. When it was most active, the ACW was associated with two e-lists of note. First was MBU-L, or Megabyte University, which was founded in 1990 by Fred Kemp and deactivated in 1997. In many respects, MBU-L was the first ongoing continuous e-list for scholars interested in computers and writing issues, thus providing an especially important community-building opportunity. For the first time, scholars on the margins of their own institutions, as well as others, had a space where they could interact with colleagues in similar situations and/or with important information and perspectives to help. Though created in 1995, while MBU-L was still active, ACW-L is generally considered to be the follow-up e-list opportunity to MBU-L. Kemp likewise created it. On ACW-L, as on MBU-L, scholars from around the world discussed computers and writing issues of interest. In 2000, when Kemp deactivated ACW-L, several e-lists emerged as possible follow-up opportunities for those who wanted to continue the conversations. Of those, TechRhet, founded by Kathy Fitch and John Walter, has been most prominent. Like its predecessors, TechRhet provides an opportunity for scholars around the world to discuss important issues. Threads of conversation in November of 2001 included the following, for instance: "Listservs and Community," "Electronic Portfolio Tools," and "MOO Research." At the same time, though, TechRhet does more than offer an e-list. It also hosts weekly MOO discussions, for instance, like the Netoric Project does with the Tuesday Café; the TechRhet discussions are weekly on Thursdays.

Beyond the e-lists specifically for the computers and writing community, it's important to note that other e-lists provide important opportunities for discussing computers and writing issues at times, just not all the time. Among these are WPA-L, an e-list for writing program administrators where technology issues come up from time to time; WAC-L, an e-list for scholars interested in writing across the curriculum, writing in the disciplines, and other like initiatives that at times include attention to technology; and WCENTER, an e-list for writing center professionals that often explores issues around online writing labs (OWLs) and other technology and technology-related subjects. On WPA-L, threads of conversation in November of 2001 included the following, for instance: "Banned

Topics—Why," "Survey on TA Training," "Books to Movies," and "Student E-mail Policies." Similarly, a recent review of the WCENTER archives reflects 12 distinct threads about OWLs in 2000 and 9 in 1999; these ranged from practical dimensions of how to build an OWL to broader discussions of how prominent OWLs are in various contexts, like high schools, community colleges, and universities around the world. In many respects, e-lists like those identified provide the most immediate and dynamic community-building opportunities for computers and writing. Across the many subjects discussed, with the various voices associated with each, e-list subscribers have the opportunity to both learn and contribute a great deal around their interest in computers and writing.

Publishing Ventures and Products

Publications also demonstrate that computers and writing is well represented as a community, no matter if such publications are print or electronic. Books are especially interesting because they bring into view two of the most important aspects of the computers and writing community: that graduate students often develop prominent scholarship at that stage of their careers and that senior scholars often specifically include graduate students and other junior scholars in their projects in order to offer them professionalization opportunities. First, scholars like Janice Walker, Todd W. Taylor, and Jan Rune Holmevik have demonstrated that graduate students can produce prominent books. Walker and Taylor developed *The Columbia Guide to Online Style* (1998) while they were completing doctoral work at the University of South Florida, and Holmevik co-edited (with Haynes) *High Wired: On the Design, Use, and Theory of Educational MOOs* (1998a), while pursuing a doctoral degree at the University of Bergen. Senior scholars like Gail E. Hawisher and Cynthia L. Selfe have provided opportunities for graduate students and other junior scholars to contribute to books they were developing. In their recent *Global Literacies and the World Wide Web* (2000), for instance, contributors included Karla Saari Kitalong and Laura Sullivan, both of whom were graduate students at the time the project began. Such professional responsibility offers a strong model for everyone developing books in computers and writing. In the two collections that I have co-edited—*Taking Flight with OWLs: Examining Electronic Writing Center Work* (2000), with Donna N. Sewell, and *Electronic Collaboration in the Humanities: Issues and Options* (2004), with Cheryl Reed and Peter Sands—I've teamed with the co-editors to follow Hawisher's, Selfe's, and other senior scholars' models and thus to make it a point to encourage submissions from graduate students and other junior scholars.

Journals in computers and writing also indicate that it is appropriately defined as a community, especially as editorial and production processes open spaces for many individuals to become involved. One of the most prominent journals in the computers and writing community is *Computers and Composition.* It was founded in 1983, as indicated earlier, and the journal is now steered by Gail E.

Hawisher and Cynthia L. Selfe, as well as a series of associate editors. One of the most telling indicators of *Computers and Composition*'s prominence is an informal tradition at the computers and writing onsite conference. When Hawisher and Selfe present the journal's annual awards (The Ellen Nold Award for Best Article in Computers and Composition, The Distinguished Book Award in Computers and Composition, and the Hugh Burns Award for Best Dissertation in Computers and Composition), they begin by inviting audience members to stand up, as descriptions apply to them. Hawisher and Selfe begin by asking editorial staff and editorial board members to stand, then scholars who have published an article or articles in *Computers and Composition,* and finally everyone who has read the journal. At the end of this activity, almost everyone in the entire room is typically standing. Interestingly, it's in the associate editor roles that one of *Computers and Composition*'s strongest features becomes evident. The associate editors come generally from the graduate programs at the University of Illinois and Michigan Technological University, where Hawisher and Selfe teach respectively, so *Computers and Composition* becomes a professionalization opportunity, helping these students learn about the editorial and production processes involved with a journal and become more actively involved in the computers and writing community. In her or his work, for example, an associate editor might review a submission, send it to editorial board members for review, receive and compile their comments, and collaborate with the editors to make a publication decision. Then, on the production side, an associate editor would interact with both Elsevier Science, which publishes the journal, and the authors, as individual issues are developed. Across all of these efforts, associate editors advance in the computers and writing community by getting to know others and at the same time learning about scholarly production.

Electronic journals also reflect an emphasis on professionalization and community. One prominent instance is *Kairos,* which Douglas Eyman and I currently co-edit. We currently publish two issues per year, each featuring a particular theme; recent themes have included "Technology and the Face of Language Arts in the K–12 Classroom" and "Disability—Demonstrated by and Mediated Through Technology." Our mission with *Kairos* is

> to offer a progressive and innovative online forum for the exploration of writing, learning, and teaching in hypertextual environments like the World Wide Web. At the same time, we hope that our balance between the cutting edge of the Web and the traditional academic need for juried publications will help electronic scholarship earn a stronger and more valued place.

Kairos was actually founded by a group of graduate students in 1996, including Mick Doherty, Elizabeth Pass, Michael J. Salvo, Jason Teague, Amy Hanson, Greg Siering, and Corey W. Wick, and it continues to incorporate students and other junior scholars in all aspects of its operation, from positions on the editorial board and staff, to special opportunities for contributions. The editorial board and staff positions offer unique opportunities as well because of the editorial review process, which features three distinct stages:

- Eyman and I receive submissions, and all staff members discuss them, deciding if the submissions should be brought to the editorial board;
- Submissions brought to the board are discussed boardwide in terms of their potential to be published in *Kairos* and whether they should be advanced in our review process; and
- Three editorial board members work directly with the authors to bring a submission toward publication.

In each of the three stages, a number of individuals, including graduate students, play key roles. Most important is the third stage; the way editorial board members work directly and openly with authors offers a unique opportunity for them all to be challenged and grow. Because our current editorial board and staff feature a diverse range of scholars interested in webtextuality, the conversations we have push us all to learn more, helping *Kairos* contribute actively to the computers and writing community's sense of webtextuality and electronic/digital publication. Our emphasis on webtextuality results from extended conversations about what it is we publish. In asking us to consider hypertexts not available on the Web for our annual awards, colleagues rightly pointed out that projects published on the World Wide Web cannot be assumed hypertextual; we remain committed to recognizing only projects freely accessible on the Web, however, so have used webtextuality as our key term since.

The computers and writing community has also been prominent in journals serving broader readerships. An important example is *Academic.Writing: Interdisciplinary Perspectives on Communication Across the Curriculum,* founded by Michael Palmquist and Luann Barnes in 1999. Each issue features a range of research projects, editorial columns, resources, and reviews focusing on communication across the curriculum, and in most cases, the sections are led by scholars in the computers and writing community—like Will Hochman, who coordinates the reviews section, Pamela Childers, who spearheads the secondary school section, and Donna Reiss, who focuses on community colleges. *Academic.Writing* also facilitates new scholars' learning about computers and writing via its editorial policies, which provide for both linear and hypertextual projects to be considered. Print and electronic journals in various fields and disciplines also feature computers and writing community members' voices. In rhetoric and composition, for instance, *College Composition and Communication, Rhetoric Review, Writing Center Journal, Pre/Text, JAC: Journal of Composition Theory, Writing Lab Newsletter,* and *The Writing Instructor* all often include projects from scholars active in computers and writing. In *JAC,* for instance, recent articles relating to computers and other technologies include the following: "Hacking Cyberspace" by David J. Gunkel (2000) and "Uploading Anticipation, Becoming-Silicon" by

Richard Doyle (2000). A number of relevant journals like *College English, Teaching English in the Two-Year College,* and *English Journal* also address English studies writ large for specific populations of readers and sometimes feature contributions from members of the computers and writing community.

Last, I would point to a range of electronic publications other than journals that demonstrate the way computers and writing is well represented as a community. An especially interesting area is hypertext publishing, an enterprise experiencing rapid growth and attracting many computers and writing community members. The most prominent hypertext publisher is Eastgate Systems, and one of the key features of its operation is the way it brings together a diverse range of authors. Computers and writing scholars like Jay David Bolter, Stuart Moulthrop, Michael Joyce, and Nancy Kaplan present their work alongside that of other talented authors like Mark Bernstein, Diane Greco, and Shelley Jackson, and the result is a publications program that challenges the way we think about the computers and writing community, compelling us to draw parallels between the work of scholars we know and those with whom we become interested. Greco's *Cyborg: Engineering the Body Electric* (1995) connects in meaningful ways with scholarship about the body in electronic spaces, for instance, and her critical examination of 20th-century writing about mind–body connections and human–machine interactions should inform any range of projects in computers and writing. I would also point to more informal publishing opportunities as important for the computers and writing community, like personal writing in various media. A number of fan fiction Web sites have been created in recent years, for example, about any range of television programs and movies, and scholars like Cynthia A. Wambeam (2000) and Christine Boese (1998) have explored their discourse and multimedia elements, providing detailed analyses of how such writing should be important to the computers and writing community.

The conferences, organizations, and publications just described well reflect why computers and writing is best represented as a community. What's perhaps all the more important is that these areas of activity have proven sustainable over time. If anything, computers and writing has grown—and grown substantially, demonstrating promise for continued development in the future. It is ultimately, then, a vibrant community, an expanding network of individuals and resources with shared histories, a shared present, and the promise of a shared future.

DEFINING THE CYBORG ERA

Because communities are situated—that is, context-specific—it's important to define them relative to constants like time. Were I to seek information about computers and writing in the mid-1980s, for instance, I would look in different places and for different features than if I was studying it in the late 1990s. Of course, the perspective being used will have as much to do with what is learned, if not more, than the time, but what the time enables is a scale, a common means of referencing specific moments in a community's growth and development.

Aside from Hawisher et al.'s *History,* which references the specific years 1979 to 1994, most scholarship visible in the computers and writing community has spoken of general ages, eras, and periods. In such cases, a name for the age, era, or period is created and intended to be distinctive. Consider the titles of three recent edited collections: *Page to Screen: Taking Literacy into the Electronic Era* by Ilana Snyder (1998); *Handbook of Literacy and Technology: Transformations in a Post-Typographic World* by David Reinking, Michael C. McKenna, Linda D. Labbo, and Ronald D. Kieffer (1998); and *Literacy Theory in the Age of the Internet* by Todd W. Taylor and Irene Ward (1998). Each title features a subtitle that forwards a time-based representation of its context—"the electronic era," "a post-typographic world," and "the age of the Internet." Other often-cited books to provide such subtitle references are the following: *Word Perfect: Literacy in the Computer Age* by Myron Tuman (1992); *Re-Imagining Computers and Composition: Teaching and Research in the Virtual Age* by Gail E. Hawisher and Paul LeBlanc (1992); *The Gutenberg Elegies: The Fate of Reading in an Electronic Age* by Sven Birkerts (1994); and *Life on the Screen: Identity in the Age of the Internet* by Sherry Turkle (1995). The books named in this paragraph do not, of course, begin to approach in number the articles and reviews that have included references to an age, era, or period in their titles and subtitles.

What interests me most about the ages, eras, and periods named is how they prove basically uninterrogated. We can't tell if Snyder and Birkerts mean the same thing by "electronic era" and "electronic age," much less the relation between all the different terms, even given the definitions the editors and authors offer in their introductions. Did the electronic era and electronic age begin with early 20th-century electronic technologies? Did the "computer age" begin with early computers like ENIAC? Similarly, did the "age of the Internet" begin with the U.S. government's development of ARPANet in the 1960s? These ages, eras, and periods seem often to reference technologies, like computers and the Internet, rather than individuals, seemingly implying that we should rely on those technologies to serve as fixed markers of time. Of course, this reality is part of the normal cycle of any technology being introduced; we first look only at the technology, then finally step back and see more. After all, relying on technologies alone as markers of time cannot realistically do so effectively because their use in that capacity reflects hegemony—intentional, unintentional, it doesn't matter. Whereas "age of the Internet" may serve as an effective marker in the United States and Western Europe, that is, it often cannot do so in underdeveloped countries; instead it may be only a level of technological maturity to be sought there. To put it simply, if we are going to reference the growth and development of any community by a technology many of its members employ, then we are most likely excluding anyone who does not have access to that technology. At the very least, we are using a term with limited significance for them, even if they still have been able to be active members of a particular community. And even in contexts where specific technologies are present, we cannot assume that the name functions any more effectively because individuals in those contexts may think and talk differently about the technologies. The introduction and use of technologies is always inherently social in nature, making any cross-contextual reference points based on those technologies problematic, even in the best of circumstances.

For contemporary computers and writing, I believe we need a new approach, one that reflects our ability to step back and see more than just technologies, and I suggest "the cyborg age" as most compelling. As demonstrated by scholars like Donna Haraway (1985), Chris Hables Gray (2001), and Anne Balsamo (1996), the *cyborg* reflects the dynamic synergy of individuals, technologies, and the contexts they share, a flexible and simultaneous emphasis that previous names for ages, eras, and periods could not provide. One of the most powerful examples of the way the cyborg reflects this emphasis comes from Allucquere Rose Stone (1995), as she recounts her experience at a lecture given by Steven Hawking:

> In an important sense, Hawking doesn't stop being Hawking at the edge of his visible body. There is the obvious physical Hawking, vividly outlined by the way our social conditioning teaches us to see a person as a person. But a serious part of Hawking extends into the box in his lap. In mirror image, a serious part of that silicon and plastic assemblage in his lap extends into him as well . . . not to mention the invisible ways, displaced in time and space, in which discourses of medical technology and their physical accretions already permeate him and us. Where *does* he stop? Where are his edges? The issues his person and his communication prostheses raise are boundary debates, borderland/*frontera* questions. (p. 5)

The term cyborg was actually coined by NASA scientists doing bioengineering experiments with mice in the 1960s to explore the possibility of engineering human body parts to enable astronauts to pursue extended space travel (Clynes & Kline, 1960), but since that time, the cyborg has been employed as a social, political, and cultural technology for reform. In Haraway's "Manifesto for Cyborgs" (1985), for instance, she reclaims the cyborg from the overly masculinist military–industrial complex and reshapes it into a basis for feminist agency. Haraway reminds us that connections are critical—whether between individuals, technologies, societies, or cultures—and helps us understand the complex and dynamic interactions among individuals, technologies, and the contexts they share.

The reasons the cyborg's simultaneous emphasis on individuals, technologies, and their contexts is so important for computers and writing in each case are threefold:

1. That individuals are foregrounded along with computers and other technologies and their shared contexts. In many scenarios involving computers and other technologies, individuals are overlooked or forgotten. It appears to matter less to actors in those scenarios that individuals have no voice and more that the latest technologies are acquired and employed, no matter the implications of those technologies. In such scenarios, individuals are only "users"—and generic,

adaptable users at that—and little or no attention is given to their unique characteristics, perspectives, and values. In this cyborg era, however, we are reminded to think carefully about these individuals and never to forget them in contexts including computers and other technologies.

2. Technologies are included, thus given an equitable place in computers and writing with individuals and the contexts around them. One of the puzzling elements of technology use is the way many individuals imagine computers and other technologies as fixed, static innovations to be adopted or resisted, rather than as fluid innovations that can be adapted. For this reason, technologies are often imagined as "right" or "wrong," a reductive binary at best. Instead, as the cyborg era as a term suggests, technologies should be foregrounded equitably with individuals and the contexts around them, enabling a more careful and reasonable assessment of those technologies, one not restricted by problematic binaries and instead featuring attention to the unique character and implications of the technologies.

3. Contexts shared by individuals and technologies are prominent too, necessary and important prominence too often neglected. At first glance, making the case that the contexts in which individuals and technologies interact prove vitally important seems almost cliché at this stage of the evolution of the computer and writing community; clearly many of us understand that contexts matter. At the same time, contexts are still too often neglected or forgotten. Computers and other technologies are adopted because other institutions are using them, for instance, or because an individual faculty member believes them to be best, rather than on the basis of a detailed assessment of their applicability for the specific contexts into which they are to be introduced. For this reason, foregrounding contexts alongside individuals and technologies is key. With the unique advantages outlined in this paragraph, the cyborg age seems a particularly apt term for the dynamic nature of the contemporary computers and writing community.

Perhaps the most important reason for adopting the cyborg era as a term, however, is what it can do in application. Put simply, a cyborg era requires agency and activism. As Haraway and other scholars have shown, the cyborg is an agent for change, never a pacified or go-along figure in any context. Using the cyborg era, then, as our term, enables us to develop and articulate specific agendas for computers and writing. We might make it our goal, for instance, to help provide and promote access to technologies in the sorts of underdeveloped countries already referenced. In such a case, we would work to create sustained opportunities for individuals in those contexts to learn about and make use of technologies, at the same time being attentive to the imperialist implications our project may have. We might want to align our goals with the sort of democratic empowerment suggested by Paolo Freire in his series of books on critical pedagogy, especially *Pedagogy of the Oppressed* (1970), and this emphasis would have us seeking to empower individuals in the countries to educate themselves on their own terms and for their own purposes, making large-scale democratic decisions along the way. One approach we might take is to provide sustained technical support. One of the critiques of initiatives like Net Day, after all, has been that computer technologies are provided, but without support or any education about how to use them, so we could learn from such critiques and develop a long-range plan. In this cyborg age of computers and writing, projects like that described can be a hallmark of our work because they feature focused attention on individuals, technologies, and their shared contexts all at once. We know we should not introduce technologies into a context without attention to the dynamics of that context and individuals present in it, like we know talking to individuals about technologies when they have no access to those technologies results in little gain for everyone involved. In this way, the cyborg era as a term extends beyond those previously suggested in scholarship.

MANY VOICES: A MONTAGE

Although I argue in this chapter and indeed throughout this book that the contemporary computers and writing community is best represented as in the cyborg era, I want to emphasize that the most powerful basis for this argument is not what I might say as an individual, but rather what many in the community have said in recent years. In this final chapter section, I present a montage of scholarly voices from the cyborg era of computers and writing community, each specifically demonstrating by subject a simultaneous emphasis on individuals, technologies, and the contexts they share. The earliest excerpts are from 1993, but most fall between 1998 and 2001. I invite readers to think with me about the many compelling voices here.

Cynthia Haynes (1998): Let us begin to invoke the literacies of our future critically, limited only by our imagination and a dialogic fantasy. Let us imagine a listserv moderated by prosthetic rhetoricians, writing email messages to phantom subscribers of the list. They all believe in the listserv. It serves the list members by its prosthetic memory, dutifully storing in its archives the amnesiacal data flung into the ether by one quick press of a button. Control X. Send message? Y or N? I don't remember what

Sarah J. Sloane (1999): In our professional writings to date about computer-mediated communication, we often forget to note the echoes of personal experience that reverberate in the ways we approach writing. We need to be more critically aware that our encounters with new communicative technologies are always colored by memory, informed by learned response, and haunted by earlier experiences with writing, reading, and communicative technologies. Further, our technologies to do here. Why are there no other options? I need my agent/emissary to make this decision. I don't remember why the listserv serves me. I have amnesia? Perhaps I am an emissary. Hmmmm. I am a prosthesis? I have replaced a list append/age with writing. Yes. Writing message to disk. (p. 90)

themselves are always haunted by their own individual and cultural genealogies. When researchers in computer-based writings explore the relations among readers, writers, texts, and technologies, a close analysis of the genealogies of each of these components is crucial. (p. 50)

Susan Romano (1993): I would hate to see the egalitarianism narrative function as a legal text— as a promise whose occasional fulfillment grants it an unexamined legitimacy. Such promises are dangerous because when they do not materialize, failure must be declared—on the part of the instructor, a particular student, or a particular mix of students. Claims for an automatic egalitarianism engendered by technology are particularly invidious because if the technology is inflexible, infallible, and ever-enabling, then human beings absorb blame for failure. Each exemplary success story suppresses alternative stories and knowledges about interactive, electronic discourse, whose authors are fearful of bringing them to light. I wish for more moderate aims for classroom happenings and for less stunning accounts of real-time conferencing. As instructor in these classrooms, I need the freedom to not measure up to utopian claims so that I can notice what it is that students do with words without trying to fit these words into a grand narrative. And egalitarianism is indeed about to qualify as a grand narrative within our discipline. I want space to think about the riddle that really does interest us and that we have not yet unlocked: not how can we affirm that technology converts classrooms into egalitarian spaces, but rather what are the multiple phenomena—positive, negative, and polyvalent— that networking facilitates or mandates? How does it alter the way we currently construct the act of writing? (p. 26)

James E. Porter (1998): Any use of the computer as a writing instrument entails both ethical and legal obligations. It is deceptively easy to articulate the general principles of computer ethics for writers and writing teachers: Writers who use and produce electronic text should adhere to ethical and legal guidelines for the use of computing facilities and resources. Writing teachers who use computers for instructional purposes are responsible for guiding their students toward (and themselves adhering to) ethical and legal use of computing facilities and resources . . . What is difficult is determining ethical action in any particular instance. (p. 22)

Bertram C. Bruce and Maureen P. Hogan (1998): We tend to think of technology as a set of tools to perform a specific function. These tools are often portrayed as mechanistic, exterior, autonomous, and concrete devices that accomplish tasks and create products. We do not generally think of them as intimately entwined with social and biological lives. But literacy technologies, such as pen and paper, index cards, computer databases, word processors, networks, e-mail, and hypertext, are also ideological tools; they are designed, accessed, interpreted, and used to further purposes that embody social values. More than mechanistic, they are organic, because they merge with our social, physical, and psychological beings. Thus, we need to look more closely at how technologies are realized in given settings. We may find that technological tools can be so embedded in the living process that their status as technologies disappears. (p. 270)

Gail E. Hawisher and Cynthia L. Selfe (2000): Certainly, the development of computer technology, while benefiting from the contributions of many peoples and countries, remains essentially flavored by exported American economic and cultural values. These complex social formations, moreover, continue to exert a great deal of influence on the literacy practices that characterize the Web today. They do not, however, comprise the whole of the story, nor do they offer satisfactory representations of the literacy practices in other cultures. (p. 5)

Dawn Rodrigues (2001): By learning what the technology could do, and by thinking about how the capabilities could enhance the teaching of writing or could suggest new ways of teaching that older technologies didn't suggest, many faculty across the country did what I did: we continually explored new tools, but always adapted them to the contexts we faced. Then, as now, we learned how to step back and re-examine the teaching and learning situation, shifting to different uses of technology as it became available and as it fit our varying approaches to instruction. *Then* it was invention software, word processing, and early bulletin board software. *Now* it is course management tools, MOOs, and collaboratively-authored Web sites. Then, as now, many of our colleagues resisted joining us in the process of designing computer-based learning environments.

Mark Warschauer (1999): Literacy is frequently viewed as a set of context-neutral, value-free skills that can be imparted to individuals. A study of history, though, shows this model of literacy to be off the mark. Rather, being literate has always depended on mastering processes that are deemed valuable in particular societies, cultures, and contexts. Changes in the technologies available for reading and writing have an important impact on how we experience and think of literacy, but technology alone is not all-powerful. Rather, technological change intersects with other social, economic, cultural, and political factors to help determine how literacy is practiced. (p. 1)

Lisa Gerrard (2000): The World Wide Web's graphical display, which allows website owners to represent visual images of themselves online, has special resonance for female websters. Like all women raised in cultures that objectify female bodies and faces, they understand that a female image on their site will be seen in terms of its sexual appeal. Webwomen who reject oppressive images of women, especially those who identify themselves as feminist, recognize that in life and virtually, a woman's value is identified with her body, and thus they are especially sensitive to graphical representations of women.

Christina Haas (1996): Until we are willing to recognize the symbiotic and systemic relationship between technology, culture, and individuals, willing to explore the implications of technology on our own literate practice and mental lives, and willing to enter fully into the various discourses of technology, scholars and teachers of literacy—arguably the group that has most at stake as technology remakes writing— are abdicating responsibility and power in helping to determine how technology and literacy are made, through use, in our culture. (p. 230)

Michele S. Shauf (2001): In a way, but only in a way, I want to examine the challenge of creating visualized arguments in electronic domains. I teach multimedia design at both the graduate and undergraduate levels at a large, public, technological university, and am constantly frustrated by what I once thought was the unwillingness of students to undertake argument in their work. When I give students a chance to invent their own multimedia projects—and I refer here to both Web sites and stand-alone CD-ROMs—they come up with a curiously limited array of topics, namely three: multimedia narratives about their families (usually their grandparents); educational multimedia for very young children (lessons in multiplication, for instance); or purely expository works laying out the facts of some topic in an old-school "who, what, where, when, and how" journalistic style . . . What I have come to believe, though, is that students in my classes conceive of such a thing as an electronic argument because they engage the computer in a language unlike mine. (pp. 33–34)

Gunther Kress (1999): Communication has always been multi-semiotic. What is happening at the moment is not in itself new; and yet it is a significant change. The cultural and political dominance of writing over the last few centuries had led to an unquestionable acceptance of that as being the case; it made the always existing facts of multi-modality invisible. The recent powerful re-emergence of the visual has, then, to be understood in that context: not as new in itself, but as new in the light of the recent history of representation, and of a nearly unshakable commonsense which had developed around that. (p. 70)

John Slatin (1998): Computers and Writing is no longer a marginal field: our longstanding concerns about the impact of technology now share center stage with questions of equity and access for students and, as tenure withers and full-time jobs disappear, for faculty as well. This change in the status of our work is bound up with fundamental shifts in the way our networked classrooms are situated within universities and colleges, and the way universities and colleges are situated vis-à-vis government and industry as well as the rest of the educational system, especially K–12. Simply put, the classroom is no longer a private space, and our networks are no longer merely local. The classroom is a node in a network connecting multiple contexts—educational, governmental, corporate, social—in which information moves in increasingly complex patterns. To change the metaphor, the walls are two-way glass: its not simply a matter of our looking out from a privileged, invisible position, but also of our being visible as never before from the erstwhile outside. We rightly fret that such visibility makes it all too easy to impose accountability measures based on inappropriate or outmoded criteria.

J. L. Lemke (1998): No one can predict the transformations of 21st-century society during the information technology revolution. We certainly cannot afford to continue teaching our students only the literacies of the mid-20th century, or even to simply lay before them the most advanced and diverse literacies of today. We must

Cynthia L. Selfe (1999c): A good English studies curriculum will educate students robustly and intellectually rather than narrowly or vocationally. It will recognize the importance of educating students to be critically informed technology scholars rather than simply expert technology users. Graduates of English studies programs will face an increasingly complex set of issues in the workplace and in the public sphere, and our failure to provide help this next generation learn to use these literacies wisely, and hope they will succeed better than we have. (p. 299)

the intellectual tools necessary to understand and cope with these issues at multiple levels signals our own inability to lead productively as professionals and as citizens. (p. 322)

PATRICIA SULLIVAN AND JAMES E. PORTER (1997): AS WE APPROACH THE STUDY OF COMPUTERS AND WRITING, WE ARE URGING RESEARCHERS NOT TO LOOK FOR THE ONE, HOLY, AND PERFECT METHODOLOGY, BUT TO EMBRACE WORKING ACROSS METHODOLOGICAL INTERFACES. WE WANT RESEARCHERS TO EXPAND CRITICALLY AND CREATIVELY THE BOUNDARIES OF WRITING TECHNOLOGIES RESEARCH. WHY? BECAUSE TO DO LESS RISKS OUR LARGER AIMS AS TEACHERS OF WRITING: THAT IS, TO HELP EMPOWER AND LIBERATE THROUGH THE ACT AND ART OF WRITING. (P. 187)

Rich Rice (2001): Whenever technology is involved there's a trickle-down effect. Some people will get the new stuff, and others will get their old stuff. That's the way it works. This will become more of an art form, however, and will be combined with "sustainable faculty development" theory. Like class-rooms whose teachers recognize that students bring with them various levels of technological literacy experience, entire institutions will begin to build on this principle. They'll do more than recognize it, however: they'll embrace chaos. Just as institutions now embrace multivocal student populations, and recognize diversity as something to celebrate, so too will they see multivocal technological literacy as something to celebrate.

Patricia Fitzsimmons-Hunter and Charles Moran (1998): It is clear that access to emerging technologies is a function of wealth, which is itself a function of the situation into which you happen to be born: your country, class, race, and gender. The more resources at your disposal, the more likely you are to have a desk and technology on or near that desk. (p. 158)

Michael J. Salvo (1999): The hypertext database of witness narratives in the Wexner Learning Center has the potential to change the way we write and structure narrative—because it changes the way history is recorded. The hypertext database is not a toy or game, but a serious (deadly serious) means of remembering. It is memory for mass culture, for a time when one voice will not suffice. . . . In the database of witness narratives, inexhaustibility is presented much more starkly and directly than can ever be achieved with the piles of shoes or the bundles of hair or reels of film shot by both sides in World War II. This is not *Schindler's List.* (pp. 283–284)

REBECCA J. RICKLY (1999): WE MUST RECOGNIZE THAT, IN THE CURRENT CLIMATE, SIMPLY WORKING WITH TECHNOLOGY DOES NOT WARRANT IMMEDIATE REWARD. THIS TYPE OF WORK CAN BE TIME-CONSUMING, DIFFICULT, FRUSTRATING, AND, FRANKLY, IT CAN LIMIT TIME THAT COULD BE SPENT ON MORE SANCTIONED PURSUITS. AND THOSE WHO EVALUATE THE WORK WE DO WITH TECHNOLOGY ARE OFTEN ILL-EQUIPPED TO DO SO. (P. 234)

Cynthia Haynes and Jan Rune Holmevik (1998b): The mission of teachers and scholars in computers and writing should be to rethink our missions with respect to technology, pedagogy, and research. We should be critical thinkers about the technologies we use in teaching writing. We should take the lead in the way technologies are shaped and constructed. One way we do that is to become developers of the technologies ourselves. Our obligation is to not reject technologies but to try out new technologies critically, and to be informed and critically oriented toward computing technologies in the classroom. Most of all, we must be willing to help improve them if at all possible. The changing purposes of education must incorporate the changing technologies with which we shape those purposes, whether we are educating ourselves about the available educational technologies, participating in the construction of new technologies, or critiquing existing or proposed new technologies.

Wayne Butler (2001): As instructors use technology more and more to automate learning and abandon the profession's traditions, theories, and practices, how long will it be before administrators, who are always looking for efficiencies and budget savings, decide all they really need to get students through required first year writing courses is standardized curricula and

David W. Chapman (1999): One often hears that the potential for the Web is great. And I would agree. The Web has the potential to sacrifice the quality of sources used by students in research for the ready availability of Web sources. It has the potential to distract students away from the analysis and reflection at the heart of a college education as they focus on the superficial appearance of documents. It has the potential to squander the precious resource of student time by focusing on the pedagogy delivered via technology, computer lab monitors, writing labs, and graders? If our profession becomes automated, standardized, and digitized, will we have a profession at all?

mechanics of Web-site production instead of on the act of writing. We may be able to avoid the Siren call of the Web as we avoided the false promises of televised classes in the 1960s and computer tutorials in the 1970s. With a little luck, the Web may never be able to reach its potential. (p. 252)

Lisa Gerrard (1993): Perhaps because there have been so few of us, until recently, and we have functioned largely outside the mainstream of English studies, people working in computers and composition have formed an unusually supportive and democratic community. I do, however, see signs that we are becoming less collegial and more divisive. These are just isolated signs—not a full-scale transformation—but they leave me to hope that, in our effort to be validated by the academy, we resist the impulse to emulate its least attractive values. (p. 24)

Alison E. Regan and John D. Zuern (2000): Networking initiatives are often guided by the claim that once all citizens have learned to read, write, turn on a computer, search the Internet, and build their own web pages, justice will have been done and we will have achieved a more nearly perfect democratic local, national, or international society. . . . We are ambivalent about such a claim. It can be a powerful tool for motivating those people who control access to cable, equipment, software, and skills, but its underlying warrant that access equals voice and technology empowers critique is groundless. (p. 189).

Beth E. Kolko (1998): The version of cyberspace allowed by current technology requires that we become more aware of how the physical world is embedded within our language. Such technologies are largely text-based; email, MOOs, MUDs, newsgroups, and other widely available forums for computer-mediated communication (CMC) are predicated on textual exchanges. While graphical interfaces for the World Wide Web are slightly changing the evolution of communication technologies and the shape of virtual space, the vast majority of online interaction still occurs with and through words. Yet despite this reliance on reading and writing, the emerging global network has rarely been characterized as a space of literacy. Nevertheless, because interaction in virtual spaces is conducted through language, learning to write [in] the world of cyberspace necessitates learning how to read the intersection of language with culture, bodies, and politics. (p. 61)

Larry Mikulecky and Jamie R. Kirkley (1998): Technology has helped transform the organization of work and literacy in the workplace. This transformation has brought with it concerns and questions about the unequal distribution of resources, the polarization of society, and the role of nationalism in a world of global employers and economies. Improved and expanded literacy is not the solution to all of the problems engendered by these transformations, but it is likely to be of help. It does seem clear that individuals who cannot function in the transformed workplace have considerably fewer choices than those who can. Our intent is not to imply that education should be solely defined by the needs of corporate America. However, as educators we have the dual goals of helping people expand their minds and their productivity as citizens. Indeed, for any long-term success to result with either goal, we must usually succeed with both goals. (p. 319)

Johndan Johnson-Eilola (1997): In hypertext, we are like angels without maps, suddenly gifted with wings discovering not only that we cannot find heaven, but also that walking made us less dizzy, that our new wings snag telephone wires and catch in door frames. We recognize the apparently radical enactment of non-linearity inherent in the node-link structure of all hypertext; we proclaim in various ways that revolutionary potential; and then we immediately rearticulate those potentials in terms of our conventional, normal practices. (pp. 13–14)

Claudine Keenan (1998): As we continue to broaden the scope of our definition of literacy to include techno literacy, we will face disappointment as we recognize that our assumptions about what students have learned in the technological arena are based on fallacy. Yes, students in the next few years will have had more experience using computers at home and in school, but how will they have used those tools? What programs are in place to prepare public school teachers for what we expect students to be able to do with computers? The electronic writing

Nancy Allen (1996): In today's workplace, many experienced workers have found e-mail to be a quick, easy way to communicate; they consider e-mail and other communication technologies to be part of the job and use them extensively. Unfortunately for newcomers to these technologies, the skills they need for using them don't come with appointment to a position. Those who are new to electronic literacy, whether in an office or a classroom, can find the initiation to be traumatic as they struggle with classroom, like its predecessor the process writing workshop, may be heading for a pedagogical blame game.

the technical and social complexities technology brings to communication. (p. 216)

CHARLES MORAN (1999): BUT THE STUDY OF TECHNOLOGY NEEDS TO BE GROUNDED IN THE MATERIAL AS WELL AS IN THE PEDAGOGICAL, CULTURAL, AND THE COGNITIVE IF IT IS TO BE INTELLECTUALLY AND ETHICALLY RESPECTABLE. WE HAVE AS A FIELD SUBSTANTIALLY EXPLORED THE WAYS IN WHICH GENDER PLAYS IN ACCESS TO TECHNOLOGY. . . . WE HAVE EVEN, I THINK TO OUR DISCREDIT, LOOKED AT THE WAYS IN WHICH POOR PEOPLE USE THE COMPUTERS THEY DO HAVE AND HAVE DECIDED THAT THEY USE THEM POORLY! BUT I WANT TO ARGUE THAT THESE ISSUES — GENDER AND TECHNOLOGY, PEDAGOGICAL USES OF TECHNOLOGY — NEED TO BE ADDRESSED IN THE CONTEXT OF THE RELATIONSHIP BETWEEN WEALTH/CLASS AND ACCESS TO TECHNOLOGY. IN THE CASE OF SOME MINORITIES IN AMERICA, WEALTH AND MINORITY STATUS ARE OVERLAPPING CATEGORIES: IF YOU ARE BLACK OR OF HISPANIC ORIGIN IN AMERICA, YOU ARE MORE LIKELY TO BE POOR THAN IF YOU ARE NOT . . . BUT THOUGH THE SUBJECTS OF THE DISTRIBUTION OF WEALTH AND OF SOCIAL CLASS SEEM TABOO IN OUR CULTURE AND IN OUT LITERATURE, AS A FIELD WE NEED TO ADDRESS THE FACT SQUARELY: COMPUTERS ARE, LIKE OTHER GOODS AND SERVICES IN OUR ECONOMY, AVAILABLE TO THOSE WITH MONEY, AND NOT AVAILABLE TO THOSE WITHOUT MONEY. (P. 206)

Tari Lin Fanderclai (2001): Obviously our goal is universal access (and by that I don't mean "everyone will have access"—I mean we'll have access any time, any place, to both community and personal Electronic Stuff). But our current model of computer use is a sort of virtual reality model, reinforced by the fiction we read and the movies we see. We're in the computer. Or we're on the computer. The computer is a place we go to. Universal access requires a shift in that mental model: we aren't going to be in the computer— the computer will be in us, or at least on us or around us. We can't break away from our desktops by breaking them into bits to stuff in our pockets, and then synching all those bits back up with the desktop at the end of the day. We need to

be our own network nodes so that we can
actually reach our electronic possessions from
anywhere at any time. (And I'm not necessarily
suggesting planting chips in our heads, though
I might be suggesting that the buttons on shirts
aren't working that hard and could surely be
made to perform more than one task.)

CONCLUSION

With computers and writing and its cyborg era defined, we can now move forward, thinking about what the terms mean for who we are and what we value. The
chapters that follow are my attempt to think about the landscape of the contemporary computers and writing community—the issues on our minds, the most
compelling challenges we face, the responsibilities we share, the most exciting
opportunities we have before us. As I hope to have demonstrated in this chapter,
I am mindful too of the community's many important voices and what we can
accomplish together, so I hope readers will join me in thinking about and acting
on the possibilities. The cyborg era of computers and writing invites our best
efforts.

Community Voices

Joe Amato

What worries you about the computers and writing community, and why does it worry you?

My primary worries about the computers and writing community have only inten-sified in the past few years, in the past year in particular. As the logic of globali-zation, despite severe economic downturn, continues pretty much unabated, and as the U.S. public domain, construed broadly, continues to show itself either inca-pable or unwilling to sustain reasoned public discourse as to real alternatives, the fall-out for a field that found its early home in the hearts and minds of composition practitioners—i.e., writing teachers, so many of whom have been women—may well be an even greater complacency about the effects of privatization on the more artistic, un-administrative possibilities and configurations new media seemed to augur only a decade ago.

Put simply, the profession is showing signs of premature graying, from the fraught adjunct ranks through to the cushy tenure tracks. Now I don't wish to den-igrate bureaucrats—which we've all become, to varying assimilative degrees, and often with good reason, for organizations will be organizations. And I'm certainly not opposed in principle to the production and consumption of geegaws. But I would suggest that both the promise new media once held—and still hold—for teaching, and the promise they once held—and still hold—for aesthetic (and unapologetically radical) innovation, are rightly premised on our capacity, how-ever impractical, for undisciplined behavior. And I'm left wondering whether, in the face of those disparities of material resources and sheer imagination that typify the postsecondary system—a small world, to be sure, and growing smaller with each successive incursion by the private sector—whether we can muster the collective will to research ourselves on a regular basis in order to advance, for a change, some advanced knowledge of our own ever-receding hairlines.

Daniel Anderson

How did you come to be active in the computers and writing community?

When I first walked into a computer classroom, I put my PC diskette into a Macintosh computer and promptly erased my entire syllabus. While frustrating, the event was the first of many small catastrophes that keep me engaged with the act of teaching with computers and in turn with the computers and writing community. I had to think fast (classes started the next day). I had to reinvent things on the fly. Erasing my syllabus made me humble and in some ways opened me to the possibilities of trial and error in teaching. Trial and error with instructional technologies somehow overlapped with experimentations in teaching. And both were welcomed in the community of computer teachers surrounding me (then and now).

I also noticed a by-product of my initial investigation of teaching with technology. I had fun. Initially, I was busy simply getting my bearings, but I soon found that experimenting with new technologies could be rewarding on a personal level. There was the sense of accomplishment that came from figuring out how something worked. There were the "aha" moments when I realized how a technology might be applied or adapted to learning. And there were the moments when the work with a technology moved beyond simple practice toward something more creative. I had found something that invigorated me. I also found that there were others who shared this feeling. It was this personal engagement that above all brought me to teach with computers. It was this common bond of invigoration that brought me the computers and writing community.

Judy Arzt

How did you come to be active in the computers and writing community?

In the early '80s I was interested in how computers would affect the nature of writing. I had already witnessed colleagues in the fields of mathematics and sciences using computers to change their teaching. I wanted to ensure that when computers entered the field of English and composition studies, as I knew they would, they would not be drill and practice and error detection. As we began using simple word processing programs in those days (e.g., *Applewriter,* etc.), it became apparent that computers as writing tools held the potential to change the way we taught writing and married well with our theories, e.g., writing as process and collaborative learning. I took a yearlong sabbatical from my full-time teaching position to read all the literature that I could and to study classrooms where computers were being used effectively to teach writing, commuting great distances to do so. The works of early pioneers such as Hugh Burns, Helen Schwartz, and Dawn Rodrigues informed my thinking. By 1985, I was well underway writing my dissertation. I had the wonderful fortune of a better than 100% return rate on surveys seeking participants, as participants solicited more participants. In the end, I cut a 24-case study to 12, for by page 500 readers were more than convinced. To add to my good fortune, my dissertation committee accepted drafts carte blanche, eager to keep current of developments in this emerging field. By 1986, I was thoroughly convinced that computers would revolutionize the way we teach. My own teaching became radically transformed, and I moved all of my classes, whatever their curricular focus, to computer settings. Since the 1980s, I have become increasingly enamored with the field and particularly the promise that Web authoring holds for radicalizing our teaching. As Jay David Bolter in *Writing Space* and theorists writing in Hawisher and Selfe's *Passions and Pedagogies* purport, Web authoring transforms communication. I am fully committed to this medium and enamored by how it invigorates students, just as word processors did in the '80s. Web authoring makes writers multi-media composers. So, though my nascence in the field of computers and writing began in the early '80s, my thinking has continued to evolve with the technology.

Anthony T. Atkins

Why do you choose to be active in the computers and writing community?

The people involved in the computers and writing community have had a profound impact on the decisions I have made concerning the ways I use computers to teach composition. As a composition teacher working in a fully computerized web-world of mice and keyboards, I grow increasingly concerned about the way I teach students to write. The community provides insight into the ways teachers and researchers can conceive of computers in online environments. As a doctoral student of rhetoric and composition, I remain active in the C&W community because it is the only group of scholars that combine composition with rhetoric from a technological perspective. As the community has grown significantly in the last 3 years, I suspect that it will continue to expand, attracting scholars from other academic disciplines who are interested in contributing to our already rich history. As programs and departments increase in size and numbers, we rely more and more on technology to aid us in our classrooms, research, publications, theories, pedagogies, and communication. C&W facilitates theoretical, pedagogical, and practical support for teachers, researchers, and graduate students who find themselves in front of a screen to teach composition. Without the scholarship that this community has created, students in webtextual environments would fail to learn the ways in which technology affects language, meaning, reality, and communication.

Wendy Warren Austin

What's the most important aspect of the computers and writing community for you, and why is it so important?

For me, the most important and wonderful aspect of the computers and writing community is the friendliness of the people who are involved in the work. The qualities of the people whom this field attracts seem to be a sense of adventure, a willingness to take risks, and a boldness to travel in uncharted territory, balanced with refreshing humility and unpretentious modesty. Even those who are "big shots" in the field do not act as if they are the only ones who know anything about this field, or that they are too good to mingle with graduate students or scholars who have yet to earn a citable, memorable name. If anything, the folks who are leaders in this field seem to feel a sense of obligation to keep their freshness by talking to *everybody,* circulating freely rather than remaining in an elite clique of the frequently published. They are REAL PEOPLE, people you want to hang out with, but who also aren't afraid to say, "wait, have you considered looking at this author or that author?" or maybe "it's not ready for an `article,' but rather a pilot study for a larger project." It makes you appreciate their honesty, but also their experience. I have been involved in this field since 1989 and, thankfully, this aspect of the computers and writing community has not changed at all and continues to enliven, encourage, and inspire me.

I think this field draws this kind of personality because computers are something you have to take risks with and experiment with, and the discipline of composition and rhetoric draws this kind of people, too. Teaching writing to first-year students requires people to be quick-thinking, easily adaptable, intuitive, and encouraging. I have found composition scholars to be much more approachable than literary scholars any day, and those in the computers and composition field even more so. If we were all to take a Myers/Briggs Type Inventory test, I would guess the majority of us would be ENFPs or INFPs. There would be a 50/50 mix of Extroverted and Introverted types, a huge percentage of N (intuitive) types, with some S (Sensing) types. More of us would be under the F (Feeling) rather than the T (Thinking) category because of the part of us that wants our students to succeed and feel good about themselves. Most of us would fall into the P (Perceiving) category, rather than J aspect (for Judging); it would be there, but not in great abundance, maybe 5 percent at best, I'd guess. Many of us love to read the papers and write comments for further revision, but more of us hate to actually give an assignment a final grade. Nothing's ever a final draft in this field, we tell ourselves.

Alexandra Babione

How did you come to be active in the computers and writing community?

I no longer remember the number of years that have passed since my first message to MBU asking for arguments to support using computers in English writing class. I believed this community could provide me the information I needed in a relatively short period of time. Although the Internet and listservs were relatively new at the time, it seemed the logical place to visit. I was immediately rewarded and have continued to find the various listservs a valuable resource. The members are genuinely concerned, passionate about their vocation, and eager to educate.

Cheryl E. Ball

How did you come to be active in the computers and writing community?

Not too many years ago, I was getting my MFA in poetry, and I didn't even know there was a computers and writing field. My, how things change! I was fortunate enough to have a mentor, Michael Keller, who had his MFA in poetry and who worked as the director of the computer lab in the English Department. He was the one who showed me what literary hypertext was; working with Michael helped me realize the possibilities of teaching and learning in online spaces. His enthusiasm spread to me. But, there were not many graduate students or faculty members at that institution who shared my technological interests, so I started to look for conferences where I could discuss my ideas with others. In 1999, I decided to attend the CCCC in Atlanta.

I had driven to Atlanta alone and knew no one else at the conference. But, one night, I decided to attend the Graduate Student Special Interest Group, where I met Doug Eyman (who had recently become co-editor of *Kairos,* with James Inman). We began to talk about *Kairos,* and Doug suggested I submit a hyper-poetry collection I had just finished to the upcoming CoverWeb on literary hypertext. When we left the SIG, Doug walked with me, and we ran into Dickie Selfe. Doug introduced me to Dickie (who I had never heard of—me, so young in the community of computers and writing!). But, I remember that night as being my introduction to the field of C&W because I realized that this community had to have the friendliest people on earth—going out of their way at a conference of 5,000 people to make me feel welcome.

John F. Barber

What worries you about the computers and writing community, and why does it worry you?

I am concerned that we will not be able to answer the question, "What's next?" As a community, what vision do we have for ourselves? "What's next?" Where should we go? What should we do? Who should we be? Not addressing such questions, articulating answers, and developing a vision to follow means our community will continue to splinter perhaps even dissolve.

The computers and writing community has a professional, annual conference. Throughout its history it has enjoyed the benefits of being a small gathering of folks eager to talk face-to-face about their work using computers to facilitate the teaching and learning of writing. Faced with dwindling budgets and still evolving travel difficulties, universities question the wisdom of sponsoring this conference. Colleagues question their ability to attend. "What's next?" Do we seek sponsorship under the umbrella of a national organization? Do we pursue connections with other conferences focusing on related areas of interest? Or do we move the conference into an electronic venue? Given either choice, the ability to meet and share ideas is important. What vision can we develop that will allow our community to continue this opportunity in the future? "What's next?"

The computers and writing community has two professional journals, one print, and the other electronic. Through their continued publication of theory and practice both have helped provide benchmarks for what constitutes quality work in our field. "What's next?" Now that hypertextual writing is commonly accepted as existing in a "native" form, how do we encourage its utilization for communication, both in our professional journals and other channels? Speaking of other channels, should we attempt to encourage their appearance in our journals? Much work is being done with computers and literacy and scholarship that would be of interest, perhaps even inspiration, to the computers and writing community. How should we attempt to connect with that work? "What's next?"

The computers and writing community is extremely democratic. Anyone may speak up, or out. Sometimes this leads to attacks on members of the community. "What's next?" Will we direct our multivocality to ends that benefit the community as a whole? Or will we relinquish the bully pulpit to those who seek its monopolization? And, as with our journals, there are other voices outside our community that may be of interest and benefit. How do we incorporate these voices into our community dialogue? How do we incorporate our voices, individually and communally, into the dialogues of others? "What's next?"

The computers and writing community is based on technological change. As new hardware and software technologies become available we seek their incorporation in our endeavors. Initially our cause was to champion the justification for increased utilization of computer technology in the teaching and learning of writing. Now, computer technology has become all but transparent in our schools and society. "What's next?" How do we utilize current technologies in new and creative ways to teach writing? How can we use emerging technologies to encourage the transportation of writing beyond our classrooms or electronic meetings? How can we use current and emerging technologies to encourage the creation of new forms of writing for new audiences? How can we demonstrate and talk about these new forms of writing in our conferences and journals? How can we share our ideas with others? "What's next?"

Finally, in building the computers and writing community we have augmented ourselves with traits and characteristics of others in order to ensure our own survival. Ideologies, methodologies, theories, practices, technologies—we have incorporated them all into what might properly be called "the cyborg." We are something new and different now. We are not what we were. Nor are we what we will have become. But what will we have become? "What's next?" What vision can we develop OF our community? What vision can we develop FOR our community? As cyborgs, how do we best proceed? Will our progress be slow and clunky, impeded by dysfunction and lack of identification among ourselves? Will we proceed quickly and sleekly, enhanced by a rhizomatic symbiosis that signifies collaboration and social construction? "What's next?"

David Barndollar

What worries you about the computers and writing community, and why does it worry you?

I am concerned that the computers and writing community has not yet figured out the most effective way to connect what it does with the work of the larger academic community of which it is a part. The most powerful thing about the possibilities afforded by electronic writing environments is their capacity to change the ways we understand all writing. Hypertexts have been around for a long time, in the form of footnotes and glosses. Multimedia texts, nonlinear narratives, and dynamic, reader-controlled works of literature also pre-date the personal computer. But we have been unable to see those prior inventions the same way we do today because we didn't have the computer to make explicit what has long been only implicit or invisible. The computers and writing world has much to offer, both to the scholarly community and the reading population more generally.

But I fear that the connection we need is not being made, because the computers and writing community insists on speaking so much to itself alone. This problem is not unique to this community; certainly most academics find themselves addressing a small audience of like-minded people much of the time. But while most specialized academic groups' importance is understood by different groups, the computers and writing world is typically considered a fringe interest, mostly concerned with new and complicated technological gadgets. In fact, the C&W world is more concerned with pedagogy, applied theory, and the limitations of its own practice than most other disciplines; but most outsiders don't know that. Perpetuating this misperception will make it more and more difficult for those with interest and expertise in the field to find a place for themselves in an academic world that sees them more and more just as technical assistants for the "real" scholars and teachers.

Eva Bednarowicz

What's the best lesson you've learned from the computers and writing community, and why is it the best?

The best lesson learned is rarely pleasant. The value of computer-mediated communication and writing instruction to me lies in the presence of the written record, the archive of the interpersonal, electronically mediated encounter. But words that persist, I soon discovered, may persist in uneasy ways, not all of them as empowering, liberatory and loving as I initially believed. I admit to being seduced by the network's capacity for rapidly interconnecting us all into chatty, egalitarian microcommunities. But one can't be a participant of these communities without soon recognizing the potential of electronic conversations to accuse, mis-construe, alienate.

In a makeshift classroom community, students who do not know each other may engage by testing the limits of expression through loaded polemic. The spirit of play that animates online encounters among some academic colleagues may not always seem appropriate to others. Though allegedly an equal-opportunity event, computer-mediated conversation can at any time become traumatic: its assumed or implied affronts become so much more unretractable since they are visibly and emphatically "on record." Sincerity alone does not guarantee benign interpretation—electronic commentary acquires a force and momentum of its own that cannot be blunted with a smile, a shrug or a pat on the back. But as uncomfortable as the transcript can read, it can provide us with documentation for reflection on our utterances as rhetoric and not just mere "chat."

My work within the computers and writing community allows me to take extended note of instances of conversational breakdown as well as track the civil conventions that evolve in redress of such charged situations. Some of these strategies are transferable to "real life"—such as a discussion that seeks to define the offending term. Some are not: a whimsical emote in a MOO cannot be reproduced in a classroom. In any case, the transcript allows me to examine closely, if not painlessly, the actual force of the words that as writing professionals we bring to the renowned practice of "conversation."

Stephen A. Bernhardt

What's the most important aspect of the computers and writing community for you, and why is it so important?

Computers are at the center of what we do as a community interested in writing. How can we be interested in writing and avoid thinking about computers, what they do to language, how they change things once language gets inside them? So much of our lives is mediated by technology—our lives are imbricated with technologies of communication, and the computer permeates everything we do.

Because computers have such wide and pervasive effects on our communication, an interest in computers and writing—membership in the community—is not restricted to academic life. When I move in and out of various workplaces, my expertise in computers and writing remains important and valued. Workplaces face many of the same difficulties we face in academia—email inundation, system incompatibilities, the difficulties of upgrades to systems and software, the challenges of communicating across distances, the challenge of figuring out how to repurpose texts from paper to screen. They are finding ways to make texts usable in online environments and struggling to find new ways of working in virtual spaces.

Similarly, families and friends and colleagues are reconstrued by technologies—with a new and different kind of closeness and accessibility afforded by computers as writing machines. Writing with computers puts us in touch, makes us accessible, and creates and sustains important communities of practice, allowing us to live across boundaries of space, time, and money. The most important single aspect of the computers and writing community is the power of the machine to create communities, so that we depend upon mediated communication in all aspects of our lives. We are very different than we would be without computers.

Kristine Blair

What scholarly project in computers and writing has been most influential for you, and why has it been so influential?

Perhaps no computers and writing project has had as much impact on me as the co-edited collection with Pam Takayoshi, *Feminist Cyberscapes: Mapping Gendered Academic Spaces.* As colleagues in Purdue's graduate program in Rhetoric, Pam and I were fortunate to have worked with Pat Sullivan, whose mentoring in the areas of computers and writing and feminist methodology continues to influence me. Similarly, we were fortunate early on to make contact with both Gail Hawisher and Cindy Selfe, frequent guest-scholars on the Purdue campus and equally significant mentors, who listened to both Pam and me discuss the potential for this project then and at a later CCCCs. In addition to the numerous scholarly and professional reasons for pursuing the project, Cindy's advice was practical: "Do the collection. You'll meet lots of great people." As I look back and think about so many of the voices in the collection—Christine Boese, Lisa Gerrard, Mary Hocks, Sibylle Gruber, Kathy Yancey, and, of course Gail, Cindy, and Pat (who wrote the foreword), I see how very true this is, as I have been continually involved with this wonderful community of feminist scholars on a range of different projects. For instance, many members of this community were also involved in Gail Hawisher and Pat Sullivan's *Women@waytoofast* ethnography of an all women's listserv; not unlike *Feminist Cyberscapes, Women@waytoofast* brought together voices that have sustained me throughout the years since the project's completion. Indeed, each year at CCCCs, Pat and Gail organize a breakfast for the original group as well as other feminist technorhetoricians, where we share professional and personal successes, and brainstorm future projects. In this sense, Cindy's prediction was right on. I have met lots of great women through these projects, and although I don't see them as often as I would like, the knowledge that I am part of such a community that began with both *Feminist Cyberscapes* and *Women@waytoofast* continues to sustain me.

David Blakesley

What's the most important aspect of the computers and writing community for you, and why is it so important?

The computers and writing community provides us with ample occasions for perpetual renewal, as teachers, writers, and scholars living and working on the edge. I like to use the digital metaphor when I think about who we are and what we do. So I say we are an F5-community, our identities just a click away from transformation. As Peter Lunenfeld has discussed in *Snap to Grid,* digital images can be fundamentally altered because they are comprised of pixels with mathematical values assigned to them, allowing us to "morph, clone, composite, filter, blur, sharpen, flip, invert, rotate, scale, squash, colorize, posterize, swirl." New technologies in our digital world provide us with the plug-ins and power tools for reconstituting, re-imagining, divesting, merging, dividing, reforming, deforming, and informing communities and selves, perpetually, at every packet of time in space.

The pleasure and prospect of morphing selves and communities is always already real-time, uncertain, and tense. Every time I buy electronic equipment I wonder about the difference between analog and digital. Technically speaking, an analog signal functions as a carrier wave sent by a source to a host, which reads the frequency and amplitude of the signal to translate it into its equivalent in video and/or audio. The key is that the signal originates from some source and the host machine then translates the signal into its analog, an analogous representation. An analog TV is, to put it in rhetorical terms, a naïve verbal realist. It constructs analogies. It says, essentially, this is like that. There are no arbitrary signifiers in the semiotics of analog systems. However, a digital signal is converted at the source into numbers, binary code, sent at high frequency to the host, where it is completely reassembled at every manifestation. Because a digital signal is comprised of a series of numbers, it is possible to manipulate it mathematically and perpetually. A digital image, for instance, is not an analogy, but a deconstruction and reconstruction, an ever renewable and pliable blob. With its attentiveness to communication on the edge of a new era, the computers and writing community (not unlike a blob, I suppose) is refreshing and refreshed, a parlor of digital selves that I find a constant source of healthy transformation.

Shelley Rodrigo Blanchard

Why do you choose to be active in the computers and writing community?

Why do I choose to be active in the computers and writing community? The People! I am constantly impressed by the work done by C&W people in their academic, public, and private lives. I think their work with technology has a direct impact on why I like them so. Working with technologies that are morphing so fast keeps people in the field open to change, accepting of change, and humbled by the inability to stay caught up. People who are never on "solid ground" are fascinating to be around. They are constantly developing new ideas that erupt from a lack of stasis; in other words these new ideas develop from constant collisions between unimagined partners. Working around, with, and through computer technologies keeps C&W people alert to new possibilities and innovative connections. Who doesn't want to be around such a vivacious group of people?

Karin Mårdsjö Blume

What's the most important aspect of the computers and writing community for you, and why is it so important?

Right now I find that the most important aspect for the computers and writing community to address is the importance of writing for different formats. When we use the computers for writing something traditional as a report or an article, or when we use for writing material for the Web, we face very different communication situations. It is the case of "making something in writing" or using the complex options of a modern computer: text, images, sounds . . . AND making all this work together. Neither the differences have been fully explored yet, nor the difficulties. This is a truly challenging task for the computers and writing community.

Chip Bruce

Why do you choose to be active in the computers and writing community?

Newcomers to a field of inquiry are often frustrated by the gap between their ordinary experience and the codified knowledge of a discipline of study. In the field of education, for example, many have trouble connecting what they know of their own learning processes, or the experiences from their own teaching with the canonical articles and theories they are given in university courses. John Dewey argued that this gap was enlarged when we reified disciplinary knowledge, viewing it as static, and constructing it as different in kind from the knowledge we gain through daily living. If instead, we could see the disciplines as representing the ongoing processes of a community of inquiry, then the conflict between personal, situated knowledge, and historically-constituted, communal knowledge becomes a problem of melding and connecting, not choosing one over the other.

The Computers and Writing community represents for me this melding of the learner and the discipline. It does that in part by the fact that, as a relatively new field of study, it does not privilege a limited or static conception of its key ideas or even of its own boundaries. Its communal knowledge bears a complex relationship with that of the individuals in the community, but it does not stand apart, promoting rigid hierarchies of knowledge. Going along with this, the social relations in the community tend to be supportive and constructive. People see themselves as valued for their own experiences and perspectives even if they are new to the community. Old-timers find that their own inquiries remain fresh because of this openness. Thus, the community processes model this melding of the learner and the discipline. At the same time, the object of study is essentially how new information and communication technologies promote the very same processes in contexts of teaching and learning, the workplace, and other social realms. This combination of serious inquiry with openness to new ideas makes the Computers and Writing community an exciting place to be.

Hugh Burns

What's the best lesson you've learned from the computers and writing community, and why is it the best?

If I have learned one lesson, it is to demonstrate, demonstrate, demonstrate.

Who could predict the scope and the magnitude of the technological changes we have witnessed, are witnessing, and will witness? Looking back from the edge of the millennium, we know that without our computers and writing community, we have begun to understand how these always emerging digital technologies—well-designed, well-wrought—create a context for sublime teaching as well as a culture for sustained learning.

Every day more and more educators join our computers and writing community. We are continuing to grow simply because every day there is something new in the virtual world, something original being digitally shared, something ready for teachers to learn and to do to become better than they were the day before.

In my day, I wondered what a mainframe computer could do for students besides offer grammar drills. So I demonstrated. It is always time to demonstrate possibilities. Just demonstrate.

Our community demonstrated that no longer was a computer's only place in a mathematics laboratory. We demonstrated that computers could blow the walls out of the language arts classroom, for no longer is the idea of a classroom static, set in space or in time. Our community has not only forecasted a new world of possibilities but also demonstrated how new hardware and new software is useful and useable. Together, computers and writing teachers have demonstrated how these digital tools change what can happen in the classroom and beyond—in work, in play, and in life.

Collaboration continues. Once I suggested investing in computers. Now I suggest investing in connected people, on their professional development, and on building community through collaborative connections. Did I mention that teachers are still the best teachers of teachers?

Only connect, then demonstrate, demonstrate, demonstrate. The joys will follow.

Nick Carbone

Why do you choose to be active in the computers and writing community?

I like writing. I like to write and I like to teach writing. It's fun. It's a neat way to know students and to know ideas and to know colleagues. Writing shapes disciplines, defines ways of knowing, turns thinking into performance; writing has done this and will continue to do this. Writing rules; writing rocks; writing matters. Writing determines reality; it forges alliances, changes minds, creates realities, destroys beliefs, revises ideals, reaffirms truths, intensifies dissent, and every other thing.

And now we're upon an age where people write more and more, in more and more different ways, with computers—machines with chips and software that network people, connect them, link them, interweave them. So all that writing does becomes intensified and also insanely different, radically new even while at its core it's fundamentally as it has always been. This community celebrates, questions, studies, teaches, challenges, and imagines all this. It's a frustrating, hard, wonderful, chaotic, promising, and exciting place to be, to be active where the present and future of writing is.

Paul Carson

What's the most important aspect of the computers and writing community for you, and why is it so important?

Five years ago, when I started teaching Composition courses in a computer lab, two out of twenty students would come to me after the first class to say that they did not even know how to turn a computer on, much less how to use all of the programs. I would try to explain to them that knowing how to use both the computer and the programs was not optional—not for my class, and not for any kind of future that they were hoping to find after college. At that time, I had to introduce most of the class to word processing, chat rooms, and the Internet.

This past fall we were watching a baseball game and somebody started to argue about Joe Torre's history. I went upstairs and asked my friend's high school age daughter to find me the statistics for Joe Torre. Without breaking out of four separate Instant Messenger chats, she opened a browser, searched for a site, found the information, and printed Torre's statistics. I was back downstairs before the argument was over, and the whole time she was writing back and forth to six different people.

She will be in college in the fall. Electronic writing for her is no more difficult than talking on the phone was for her parents when they were her age. This is the space in which she is writing, and therefore it is the space in which we as teachers of writing have to engage her a student. Being part of the computers and writing community is no longer an option. Our students are already there.

Joanna Castner

What scholarly project in computers and writing has been most influential for you, and why has it been so influential?

Christina Haas's exploration of what she calls "The Technology Question" in *Writing Technology: Studies on the Materiality of Literacy* has been enormously generative for me. The Technology Question is: "What does it mean for technology to become material? That is, what is the effect of writing and other material literacy technologies on human thinking and culture?" What I found particularly interesting about her exploration of this question was her integration of the actual physical body, as distinguished from the more often discussed (at least among computers and writing folk) culturally constructed body. Her project has helped me rethink all kinds of activities related to writing and teaching and learning in terms of the mind-body problem. And when these things are related to specific cultures and contexts and technologies, there are lists of fascinating new projects possible.

Pamela Childers

How did you come to be active in the computers and writing community?

I think I began my involvement in this community after the administration at Red Bank Regional High School in Little Silver, New Jersey, gave me an Osborne 1 computer for the writing center in the early 1980s. I knew that I had to learn how to use the computer as a tool for writing and learning, so I had to learn WordStar and the other CP/M programs that came with the Osborne. I attended classes on the Apple 2e as well with colleagues, sessions at NCTE and CCCC, and learned from students, too. I also began taking courses towards another graduate degree that involved computers and writing. In 1985 Cindy Selfe nominated me for the position of Treasurer of ACE (Assembly on Computers in English). I became "Treasurer for Life." About the same time, I was selected a WWNFF (Woodrow Wilson National Fellowship Foundation) Fellow to study Using the Computer as a Tool to Teach Writing at Drew University. For the next two summers, I team taught this WWNFF course at SUNY–Purchase to secondary school teachers from New York City. Those were great learning experiences that pushed me to ask more questions of my ACE colleagues. I have continued to attend sessions at national conferences to learn more, helped facilitate workshops with my ACE and IWCA peers, and done some case study research that I published. Over the years, people in ACE like Stephen Marcus, William Wresch, Hugh Burns, Cindy Selfe, Michael Day, and many others have been great mentors for my use of computers and writing.

Kate Coffield

How did you come to be active in the computers and writing community?

In Fall 1993, our new WPA decided that the English department needed to get digital, so he obtained a grant for hardware. The machines arrived over semester break, and we emptied a seminar room, but there were matters of furniture and wiring. Ancient tables and chairs were salvaged from the university stores, so that the room looked like a shared backstage prop area for *Star Wars* and *Oliver!* It even had a balcony and a fireplace. Extension cords and multiple outlets were strung everywhere (once a student dropped her bag and knocked out power to half the stations). My Spring 1994 fycomp and research sections were the first to be held in our "computer classroom." Activities were rather limited with only two programs: *Word 5* and *Typing Tutor.* But stray cats would occasionally wander in, and once a cockroach crawled out of a floppy drive, which helped to engage the students (works in traditional classrooms as well).

Meanwhile, during an early Lynx voyage, I learned of the Tenth Computers and Writing Conference. There in Columbia, MO, were rooms full of folks doing hypermultimedia object-oriented somethings with fifth graders while reminiscing, "Hey, remember back in 1984 when we were considering word processing in the comp classes?" I trudged sheepishly from session to session, muttering that I'd just *arrived* in 1984 in 1994. On the last day, I met Dickie Selfe and somehow found myself agreeing to be part of his workshop at the next year's conference in El Paso. Then I joined MBU-L (Megabyte University), where the first thread was about whether we should use emoticons or toilet paper—I forget which—because Shakespeare didn't. By 1995, having given my first C&W presentation and attended the Selfe-Hawisher summer workshop in Houghton, I'd almost caught up to other people's 1990. Or was it 1890? Witnesses will attest that I later sent a book chapter draft by copy-pasting it into UNIX mail. . . .

Dagmar Stuehrk Corrigan

**What worries you about the computers and writing
community, and why does it worry you?**

The concern I have about the computers and writing community is the trepidation
by which other disciplines, namely those in the field of Education, are welcomed
or perceived by our community. Recently in the pursuit of my doctoral degree,
I've taken courses in educational technology. I find the learning theories that I
have been exposed to and the research about how people learn with computers to
be complementary to composition theory. These "new-to-me" theories have
informed and energized my research and my teaching. I find that by not looking
beyond the borders of composition theory and/or failing to acknowledge and
value what is going on in other related disciplines, the computers and writing
community is missing opportunities for furthering the field.

Eric Crump

What's the best lesson you've learned from the computers and writing community, and why is it the best?

One of the most useful things I've learned from the computers and writing community is the limits of innovation. Yep, it's a community with a history of and well-earned pride in its ability to innovate, technically and pedagogically. It's a community with a tradition of making good things happen with or without adequate resources. And yet it's that very tradition of innovation that illuminated, for me, the tether we're on. We cannot innovate indefinitely along whatever trajectories we happen to instigate. We can explore new technologies (acceptable). We can try new approaches suggested by new technologies (acceptable). We can discover that new technologies and pedagogies render obsolete certain features of the institution, like grades, classrooms, semesters, etc. (unacceptable). We can envision new kinds of learning environments (unacceptable). We can dream up new shapes for educational institutions (unacceptable).

Well, I suppose it's nice to know where the boundaries are!

Possibly I'm the last person to figure this out. And maybe it's too obvious to mention for most folks. Certainly no one seems terribly concerned about it. There are many issues and difficult challenges facing the computers and writing community, but from what I can tell they all fall within the boundary of acceptability. The challenges often involve acquiring recognition for innovative pedagogies and scholarship, but *within the realm of acceptability.*

Nothing wrong with this at all. It's the way things work. Learning this lesson helped me figure out which side of the boundary I wanted to be on, which side I *belong* on. That's an important lesson, and I'm grateful to the computers and writing community for helping me learn it!

Karen D'Agostino

What's the most important aspect of the computers and writing community for you, and why is it so important?

Although I have led my college's computers and writing initiative since 1986, the demands of my work locally have caused me to step in and out of the larger (national/international) computers and writing community. During my initial involvement with in this group in its early years, I found it to be a supportive and accessible community, sustained by the very technology that captured the imaginations of so many of us. As a member (albeit a fairly quiet one) of Megabyte University, a listserv, I found validation and support for integrating technology in the teaching of composition through this group. The Computers and Writing conferences in Indiana, Ann Arbor, Minneapolis, Logan, El Paso, Hawaii, and elsewhere felt very accessible and intimate, especially in contrast with the enormously overwhelming 4Cs conferences. (As this community emerged, it did so with a sense of humor, as evident in an early Computers and Writing conference that met in Minneapolis but was referred to "The Pittsburgh Conference.") The participants and presenters at the Computers and Writing conferences were open and welcoming. I could arrive in a state far from home and always make connections with a group of colleagues, many of whom I could count on seeing year after year. They spoke my language (technology) and helped me to expand my thinking about technology and its possibilities for teaching and learning. There were always new ideas and sources of inspiration.

As demands on my time at the community college where I work increased, I became less involved in the larger computer and writing movement. I was overburdened with a half-time administrative position coordinating our Teaching, Learning, and Technology Center, finishing my dissertation, teaching, and expanding our local computers and writing initiative to encompass six classrooms and a lab. Through my participation on the Instructional Technology Committee of NCTE, I was able to sustain some of these professional connections, yet I missed the Computers and Writing community and the stimulation and support found there.

Recently, I decided to re-immerse myself in the computers and writing community, and when I logged in the Assembly for Computers and Writing Web site, it was like coming home. It was a joyful experience to note many familiar, as well as some new names among those still working hard, pushing the envelope supporting and challenging one another. It is incredible to be part of such a vital group. The Computers and Writing community has encouraged, supported, and sustained me, and now, upon my return to it has re-embraced me. I feel very fortunate to part of this special community of educators and look forward to moving back into a more participatory role. We may all now be part of "The Cyborg Era"; however, the computers and writing community has always been remarkably human.

Michael Day

How did you come to be active in the computers and writing community?

In about 1987, as a second semester grad student in U.C. Berkeley's rhetoric program, I was still cutting and pasting typed drafts of my term papers, tearing my hair out as I did so. I recall someone suggesting I should try a computer, so I rented one and it changed my world. My first computer was a Mac Plus (I still have it!). At about the same time, my mentor Seymour Chatman announced in the composition pedagogy class for TAs that he'd had a request from Joe Williams to help find someone willing to computerize the exercise from Williams' well-known book *Style: Ten Lessons in Clarity and Grace.* Chatman also wanted someone to investigate how we might use computers in teaching writing. I volunteered, and was soon matched with Berkeley's Instructional Technology Program (ITP), where I worked with Owen McGrath on the style exercise program (Stylex) for Williams' book. I was invited to teach in a microcomputer lab in the computing center in about 1988, but soon thereafter the English department got interested in having its own lab. I advised on the design of this lab and a lab in the study skills center, then was hired by ITP in 1990 to be their first head of the Writing Focus Group (WFG). I consulted with writing instructors on how they might use computers and the Internet in their classes, demonstrated software, and held monthly meetings of the WFG with guest speakers such as Fred Kemp, Lisa Gerrard, Irene Clark, and Ann Watters.

ITP generously sent me to the Computers and College Writing conference at the World Trade Center in 1990 and to the Computers and Writing Conference in Biloxi, 1991. It was at these conferences that I became aware of the international community of computers and writing scholars and discovered discussion lists such as Purtopoi, Wcenter, WPA, and MegaByte university. On these lists I found community and a support system for what would become my primary research field. Finding this community was extremely important to my professional growth, for most of my colleagues at Berkeley were not that interested in teaching writing; I was feeling isolated and in danger of giving up. From the discussions on these lists developed conference presentations workshops, publications, teaching exchanges and much more. People sometimes ask how I find the time to present and publish so much; I answer that this computers and writing community has a great way of dragging people into activities that are interesting and fun, so it's natural for us to want to share in print and at conferences. Without the support of the community, I could not have published Stylex in 1996, chaired the NCTE Assembly on Computers in English from 1996–1998, and hosted the 15th Computers and Writing Conference in Rapid City in 1999. The computers and writing community is truly a rhizomatic groupuscle, a hivemind engaged in collaborventions, a home for weary writing teachers, and I am proud to be part of it.

2

Computers and Writing
From 1960 to 1979:
Cyborg History

Computers and the Teaching of Writing in American Higher Education, 1979–1994: A History, written by Gail E. Hawisher, Paul LeBlanc, Charles Moran, and Cynthia L. Selfe (1996), has often been read as the most prominent history of computers and writing, although the authors seemingly did not mean it to be read as such. Indeed Hawisher, LeBlanc, Moran, and Selfe carefully qualify their project, acknowledging that other histories might go back further in time or go beyond American higher education. In the margins of *History,* for instance, the authors provide important technological advancements and life events of the 1960s and 1970s, and in her preface to the book, Lisa Gerrard points to K–12 innovations of those decades as especially valuable to know. Beyond the specific focus of *History* and how it might be expanded, I also note that the book's defining computers and writing to be a subfield of composition studies represents a limitation as well. A community view that spans many fields and disciplines invites much broader histories.

In this cyborg era, we have an opportunity to look critically and carefully at what we know about the history of the computers and writing community and to expand it in meaningful ways, as appropriate. To accomplish this goal, we might utilize what could be termed *cyborg history.* As a concept, cyborg history may be understood as history that innately questions itself and other histories in order to provide an alternate view. Cyborg history simultaneously emphasizes individuals, technologies, and their shared contexts, so for the purposes of this chapter, it compels a look at the entire lives of individuals and technologies, not just 1979 and post-1979 events of note. Additionally, because cyborg history requires a critical interrogation of existing histories, it invites us to ask questions about what we know of computers and writing thus far and what else we should or might know. In this way, cyborg history does not write an alternate history itself per se, although its effectiveness might often spur the writing of such alternate histories; instead it is a means of deciding what information has been included and excluded in histories, thus in public memories. Because it is defined this way, cyborg history is especially well suited to identify problematic social issues, as well as those cultural and political.

The advantages of applying the concept of cyborg history to accounts of the computers and writing community become most powerful when specific individuals are involved. Imagine the case of Rhonda Taylor, a 57-year-old, African-American, middle-school English teacher who regularly employs computer technology in her classroom and as a result has been recognized as a technology pioneer in her school district. Rhonda was born in 1946, so she graduated from high school in 1964. Rhonda attended the University of Michigan, where she earned a bachelor's degree in English education in 1968, and she returned for a master's degree in educational administration, graduating in 1977. In both programs, she worked extensively with technologies for teaching, especially audio recording and playback devices, and she has attended almost every computers and writing onsite conference, including the first in 1983. Would it be fair to include only 1979 and post-1979 experiences in histories associated with who she is in the computers and writing community? Of course not. Clearly any number of pre-1979 experiences must influence who she is today, both as a community member and more generally as a teacher and person. But that's in essence what uncritically adopting only a 1979-to-present-only history of computers and writing alone means, whether intentional or not. The implications are that important.

In this chapter, I articulate and explore key questions that should inform expanded histories of the computers and writing community:

- What about technologies other than the computer?
- What about resistance to technologies?
- What about the influence of women?
- What about the influence of minorities?

These questions emerge from cyborg history's emphasis on questioning previous histories, as well as its simultaneous emphasis on individuals, technologies, and their shared contexts. In considering each of the four questions, I emphasize the years 1960 to 1979, as I believe the large-scale social and cultural transformations of that era make it especially compelling to be included in any computers and writing history, given prominent contemporary attention to access and diversity issues. Additionally, these questions are important because visions of technology innovation have proven too often White and male. Ordinarily I might be comfortable with blanket categories on, say, the influence of women or the influence of minorities, so I add them here for their specific political value. Ultimately, I do not suggest that a 1960–1979 timeline represents the only viable means by which our understanding of the computers and writing community's history can be expanded; on the contrary, I believe the best-case scenario will occur when many histories are being crafted and considered.

WHAT ABOUT TECHNOLOGIES OTHER THAN THE COMPUTER?

The concept of cyborg history compels us to ask why computers are often the only technologies in accounts of the computers and writing community's history. Any history of computers and writing should seemingly address the wide range of advanced technologies evident when computers themselves were being developed, after all. Such technologies include those in classrooms, as well as in the space program and in military arsenals. Additionally, science fiction's portrayal of technologies in the 1960s and 1970s proves especially important. And all of the technologies and visions of technologies together contribute to helping us begin to think through the critical implications of technology use. Computers and writing histories must address these technologies and technology issues because they shape at least in part how individuals think about the many technologies in their lives today, including but not limited to computers.

We might begin with classroom technologies utilized in the 1960s and 1970s, as these important technologies are too often forgotten in computers and writing histories. The term *educational technology* was actually introduced in 1948 by W. W. Charters, and *instructional technology* first appeared in 1963, coined by James Finn. Robert Glaser introduced a related term, *instructional systems,* in 1962. Building on earlier efforts as far back as the 1920s, classroom innovations with television and video received considerable attention for possible classroom applications in the 1960s and 1970s, especially as they contributed to emerging educational systems. By the 1960s, for instance, the American Department of Audiovisual Instruction (DAVI) had published a series of books on such technologies and their educational applications, including *Teaching Machines and Programmed Learning: A Source Book* (Lumsdaine & Glaser, 1960). This book, in particular, drew on work by psychologist B. F. Skinner, who was developing a self-paced learning technology designed to offer users immediate feedback about errors, thus helping them correct those errors. Another DAVI book of note was *The Changing Role of the Audiovisual Process in Education: A Definition and a Glossary of Related Terms,* published in 1963 (Ely). Reflecting rapid growth in technology development and its applications for teaching and learning, DAVI changed its name to the Association for Educational Communications and Technology (AECT) in 1971. At the same time I describe important classroom technologies of the 1960s and 1970s, I should also indicate that their use was not widespread, when the enormity of the educational enterprise at all levels is considered. In *Teachers and Machines: The Classroom Use of Technology Since 1920,* Larry Cuban (1986) points to several contributing factors, including specific technologies' incompatibility with teacher goals and expectations and the often nondemocratic technology adoption decision processes utilized in school systems.

In the 1960s and 1970s, space-race technologies were prominent in the mass media and in the world in general, so it follows that these technologies either directly or indirectly influenced individuals. Such prominence proves valuable in terms of computers and writing histories because it reflects the sort of public engagement of technologies that has continued since and because many who are and have been active in the computers and writing community lived through those decades and experienced the prominence of space-race technologies then. As early as 1961, President John F. Kennedy made the space race a public issue by openly stating that the United States would be the first country to have astronauts land successfully on the moon. Although John M. Logsdon (1970) and other historians have tried to soften Kennedy's statement in historical accounts as a carefully conceived public relations move, it nonetheless was a powerful challenge for the United States and the world at that time and a powerful public endorsement of technological innovation. Throughout the 1960s, the space race became essentially one of the world's most exciting dramas, with Kennedy's challenge finally being realized by NASA's Apollo 11 project, which saw Neil Armstrong and Edwin Aldrin walk on the moon on July 20, 1969. From the contemporary perspective, Kennedy's public endorsement of the space race seems especially telling because Presidents George W. Bush and William Clinton and other key world leaders have themselves prominently endorsed computers and the Internet in the late 20th and early 21st century, showing a direct link between the space race era and today. In 1998, for instance, Clinton gave the commencement address at the Massachusetts Institute of Technology (MIT), and he offered challenges for the graduates, emphasizing the need for technological innovation to be socially responsible and for technologists to play a key role in that effort. This address, along with others and with governmental initiatives like Goals 2000, has generated considerable popular attention for technologies, demonstrating their prominence in the contemporary world.

In the 1960s and 1970s, military technologies threatened many individuals around the world, and this very real threat influenced many present-day computers and writing community members, whether they were adults or children at the time. World governments' use of military technologies in those decades taught a simple but powerful lesson: No one is safely out of the reach of technologies, even if we are in our homes and our country is powerful. Beginning with the Cuban Missile Crisis in 1962, the Cold War indeed brought stark reality to the growing potential of rival nations to threaten each other. The Crisis itself, in which American spy planes discovered launch sites for nuclear weapons in Cuba, shows how individuals could not avoid the impact of military technologies, as from Cuba, missile strikes could have been made on many different fronts in the United States. Since that time, the range and power of such military technologies have only increased, making the Cuban Missile Crisis a particularly important event—perhaps a landmark event—in terms of how such technologies influence who we are today and how we think about other countries in the world. Only a few years later, the Vietnam War began, and this event demonstrated clearly that those countries with the most significant and sophisticated arsenals were not always clearly superior in battle. Such a perspective itself was not new, of course, as the United States itself was the winner of such an uneven battle in the 18th century in winning its independence from England, but it was the first example in the modern era, when individuals had assumed advances in military technologies were more powerful proportionally than those in colonial America. In making the connection between such military technologies and the computers and writing community, I do not imply that a threat like a computer virus somehow equates to a nuclear missile's being in range of our homes; that would be naive at best. At the same time, movies like *Wargames,* which suggested that a computer had the power to initiate and fight a global nuclear war, have made the connections more tangible, more easy to imagine, especially for individuals not familiar with computers and other contemporary technologies. We make a mistake if we fail to account for military technologies in histories of computers and writing that attempt to include the way individuals sometimes imagine technologies as threatening, as we may not understand fully the serious nature of those reservations without such a parallel.

At the same time classroom, space race, and military technologies were influencing a wide range of individuals in the 1960s and 1970s, portrayals of future technologies were also prominent in science fiction television and literature. These portrayals shaped not just the way individuals thought about technologies at that time, but the way they imagined their own futures amidst technologies and the way they imagined the futures of the world around them. In this way, science fiction in those decades played a significant role in the lives of many who choose today to be active in the computers and writing community, as well as those who do not. Popular television series in the 1960s and 1970s included *The Twilight Zone, Lost in Space,* and *Star Trek,* and these offered viewers fantastic voyages to both alternate realities and other worlds. Technology was positioned often as the means by which such realities and worlds could be reached; on *Star Trek,* for instance, the U.S.S. Enterprise took Captain James T. Kirk and his crew to other planets at warp speed. Technology was not always perfect—the Enterprise was known to break down and need repairs from Scotty, a character who always managed to fix the ship just in time—but it never let down the crew in a way that cost them their lives or sacrificed their mission. In literature, Frank Herbert's *Dune* (1965) likewise took readers to other planets and a future full of starships, imaginative creatures, and interesting landscapes. Other authors particularly active in the 1960s and 1970s like Isaac Asimov and Daniel Keyes created similar possibilities in their books and short stories. Beyond the way science fiction in those decades shaped how individuals might imagine their futures and those of the worlds around them, it also influenced the way they would interact with future science fiction, especially future visions of technologies, and this reality is important as well for a history of the computers and writing community because it shows that we have not just begun to come to science fiction with cyberpunk novels and other contemporary television programming and literature. Indeed many in the computers and writing community have long thought about and interacted with science fiction, thus science fiction's visions of technologies.

WHAT ABOUT RESISTANCE TO TECHNOLOGIES?

Cyborg history next compels us to ask why resistance to technologies is only rarely included in accounts of the computers and writing community's history. In the 1960s and 1970s, resistance to technological progress emerged in the discourse of a range of scholars, like Eldridge Cleaver, Jacques Ellul, Martin Heidegger, and Herbert Marcuse. For the computers and writing community, such resistance is particularly important because it reinforces the value of thoughtful and well-articulated resistance over reactionary and ill-conceived responses to computers and other technologies. If we can think about resistance as an important element of computers and writing histories, then we reinforce our collective investment in the careful and responsible use of technologies.

One of the immediate problems with discussions of resistance is in determining how it should actually be defined. In this pursuit, it's useful to examine specific instances of 1960s and 1970s resistance to technologies, as these decades featured prominent scholarship about the critical implications of technologies. Such perspectives should be included more often in histories of the computers and writing community not just because they still prove telling, even when applied to 21st century technologies, but because they ground new perspectives, new approaches to resistance. A particularly interesting case emerges from the relationship of philosophers Martin Heidegger and Herbert Marcuse. In Douglas Kellner's introduction to Marcuse's *One-Dimensional Man* (1964), he explains that Heidegger and Marcuse began studying together in the 1920s and the early 1930s. Between the World Wars, as described by Kellner, Marcuse disassociated with Heidegger: ". . . he [Marcuse] later broke with Heidegger after the rise of National Socialism in Germany and Heidegger's affiliation with the Nazi party . . ." (p. xiii). This schism between the thinkers demonstrates my reason for emphasizing them in this section. It is the Nazi warfare and social technologies that sees the two philosophers split, and given this chapter's interest in reading the social and cultural transformations of the 1960s and 1970s into histories of the computers and writing community, the forms of resistance surrounding technologies seem vitally important to consider. Heidegger and Marcuse offer much of value in their well-developed and well-crafted perspectives.

Heidegger's definition of resistance is developed in his extended discussions of the nature of technology, and in his emphasis on humanity, as well as in critiques of his approach, we can see how his views relate well to the computers and writing community. One of Heidegger's most widely cited essays is "The Question Concerning Technology," which appears, logically, in *The Question Concerning Technology and Other Essays* (1977). In *The Question,* Heidegger takes an interesting tack in presenting technology, defining it, first of all, as a mode of human activity, instead of as a mechanical innovation. In this pursuit, he further outlines how humans take an "instrumental" perspective in interpreting the world and thus are trapped or determined by the relationship of the technologies they employ to what he calls "destining." Heidegger's principal interest is in discussing how individuals can pursue freedom from technologies, a state or result that may not be possible. For many of us in the contemporary computers and writing community, the idea that technology shapes our lives is likely not frightening—we often choose that path intentionally, after all—but for anyone not invested in technologies, such a reality would indeed be uncomfortable. Some scholars have found grounds upon which to challenge Heidegger's sense of technological determinism, and these challenges help clarify why Heidegger's views prove so important for computers and writing histories. Andrew Feenberg (1999) attempts to identify problems inherent in Heidegger's view, for instance: ". . . [Heidegger] warns us that the essence of technology is nothing technological, that is to say, technology cannot be understood through its functionality, but only through our specifically technological engagement with the world. But is that engagement merely an attitude or is it embedded in the actual design of modern technological devices?" In this excerpt, Feenberg (1999) challenges Heidegger's thinking in an insightful direction, choosing to ask what human technologies look like in everyday lives, instead of engaging their philosophical nature. Feenberg also believes that Heidegger essentializes technology and that the essentializing does not allow for the diverse range of technological innovations that individuals might encounter in the world: "Unfortunately, Heidegger's argument is developed at such a high level of abstraction he literally cannot discriminate between electricity and atom bombs, agricultural techniques and the Holocaust." As Feenberg's responses to Heidegger show, resistance can and should be questioned, if it is well-developed; such questioning, like the resistance itself, serves to illuminate perspectives more clearly.

Herbert Marcuse also develops thoughtful resistance to technology, the sort of critical perspective we should include more often in histories of the computers and writing community. In *One-Dimensional Man,* Marcuse (1964) critiques capitalist ideologies, especially as they influence industry, politics, and technology. He situates the technological progress of war and industry with a deterioration of human values, a perspective seemingly directly emerging from his flight from the German Nazi state that Heidegger embraced:

Today, in the prosperous warfare and welfare states, the human qualities of a pacified existence seem asocial and unpatriotic—qualities such as the refusal of all
toughness, togetherness, and brutality; disobedience to the tyranny of the majority;
profession of fear and weakness . . . ; a sensitive intelligence sickened by that which
is being perpetrated; the commitment to the feeble and ridiculed actions of protest
and refusal. (p. 243)

In this excerpt, Marcuse accounts for the dehumanizing influence of technological
advancement: the very human qualities that should be most social and patriotic
are not seen as so and instead are seen even as harmful to the welfare of individual
societies. What Marcuse's reading of human values against ideological influences
in technological states establishes is that no easy answers existed for engaging and
transforming problematic social structures. Because we now know the way technologies have evolved and the value of diversity and multiculturalism, we can
make sense of complex social, cultural, and technological factors that were prominent in the 1960s and 1970s, but Marcuse's scholarship shows that negotiating all
of the factors at that time was extraordinarily challenging. And the disorientation
and frustration individuals may have felt then does not simply go away today, as
similar critical issues relating to a host of new or more technologies in the computers and writing community have emerged. Marcuse (1964) also writes,

By virtue of the way it has organized its technological base, contemporary industrial
society tends to be totalitarian. For "totalitarian" is not only a terroristic political
coordination of society, but also a non-terroristic economic–technical coordination
which operates through the manipulation of needs by vested interests. It thus pre-
cludes the emergence of an effective opposition against the whole. (p. 3)

In this excerpt, it is evident how, for Marcuse, technological progress determines
the social, political, and economic fates of industrial societies. In light of this possible totalitarianism, it is important to note that the emergence of technologies in
the 1960s and 1970s happened on a large scale; it is not as though individuals in
that era could simply have chosen not to participate in the societies in which they
resided. Participation meant utilizing technologies that were naturalized into societies, as it does today, so in this way, citizenship required what Marcuse terms a
"passive" approach to any technologies. Individuals could not actively challenge
the technologies in their lives because, without the technologies, they would not
have been able to function within their capitalist states. Sadly, either direction—
whether to adopt and use the technologies available or to resist them—involved
making substantial sacrifices, and these are the same problematic options many
people face in light of similar influences today.

WHAT ABOUT THE INFLUENCE OF WOMEN?

Cyborg history calls on us to wonder aloud why women's diverse contributions
are not recognized more often in histories of computers and writing. Women like

Grace Murray Hopper and Evelyn Boyd Granville were among the early pioneers of computers. More, the way women contributed to and shaped the social and cultural transformation of the 1960s and 1970s clearly influences contemporary attention to access and diversity in the computers and writing community. This influence is perhaps best seen in the breadth of women's perspectives and experiences, from feminist scholarship and reform in the academy to feminist voices from outside the academy. Reading all of these contributions into computers and writing offers much broader and more realistic views of the community's history that often have been evident before.

Women have contributed a great deal to the development of computers in the 20th century, and the 1960s and 1970s were no exceptions. These innovators and their contributions should be included more often in histories of the computers and writing community because they establish that women developed many technologies from which we all have benefited and continue to benefit. Grace Murray Hopper, who served as an officer in the U.S. Navy, was a key contributor to the design team that introduced the programming language COBOL in 1960. She helped program the Harvard Mark I and Harvard Mark II computers and developed the first computer compiler. Hopper was later recognized with the National Medal of Technology; she was the first woman to win it. Hopper additionally is known is for discovering the first computer "bug," literally a moth wedged between relays and recorded by the Mark II computer.

Another key innovator in the 1960s and 1970s was Evelyn Boyd Granville, who was one of the first African-American women to earn a PhD, hers in mathematics from Yale University. Granville worked with the National Aeronautical and Space Administration (NASA) and developed computer technologies to enable trajectory systems for the Vanguard, Mercury, and Apollo projects. In the 1960s and 1970s, Alexandra Illmer Forsythe, another important figure, authored computer science textbooks, including the first ever textbook in that field, *Computer Science: A First Course* (1969). She later authored a second edition, which was released in 1975, and co-authored another book (with E. I. Organick), *Programming Language Structures,* published in 1978. Another woman of note, Erna Schneider Hoover, after teaching for a number of years at Swarthmore College, became a researcher at Bell Laboratories. While there, she invented the first computerized switching system for telephone traffic; she was awarded a patent for this innovation in 1971, and its principles continue to guide and ground contemporary switching systems.

Although technological innovators like Hopper and Granville prove important to histories of computers and writing, a number of other women of the 1960s and 1970s also played a key role. Scholars associated with the rise of second-wave feminism, for instance, began to etch a more complex and sophisticated definition of gender, emphasizing the social, cultural, political, and historical alongside the biological, psychological, and sexual. Such a multifaceted approach to gender grounds many contemporary projects in the computers and writing community,

like Kristine Blair and Pamela Takayoshi's *Feminist Cyberscapes: Mapping Gendered Academic Spaces* (1999) and often-cited articles by Susan Herring. Herring,
in particular, with her scholarly investigations of gender in electronic discourse
(1993, 1996, 1999), helps us think about the importance of second-wave feminism
in the 1960s and 1970s because scholars in those decades also studied gender in
discourse. In many respects, then, second-wave feminism provides not just an
intellectual foundation for the contemporary computers and writing community,
but also a series of models for how scholars could forge links between large-scale
social and cultural change and their individual research programs. Because the
1960s and 1970s were not long ago, we also do not have to point to scholarship
alone in demonstrating why second-wave feminism is important to include in
computers and writing histories. Senior scholars today—women and men alike—
lived through those decades, thus must have been influenced by them and their
attention to definitions of gender. A contemporary scholar as young as 50 would
have been in her or his teens in the 1960s and in his or her 20s in the 1970s. In this
way, we continue to see the influence of second-wave feminism in the computers
and writing community.

Another important way the 1960s and 1970s contributed to histories of computers and writing is by locating women's social and cultural advocacy not just
with feminist scholars in the academy, but with many women in the world. These
connections remind us that although we work in the academy or industry and
interact with only a limited number of individuals on a day-to-day basis, our
efforts have much broader implications. Betty Freidan's *The Feminine Mystique*
(1963) portrays key reforms not just between women and the changing American
society around them, but also among women. In her opening chapter, indeed in the
first paragraph of the book, Freidan (1963) demonstrates how feminist politics can
affect women in suburbia, a site not as often connected with calls for reform:

> It was a strange stirring, a sense of dissatisfaction, a yearning that women suffered in
> the middle of the twentieth century in the United States. Each suburban wife strug-
> gled with it alone . . . [a]s she made the beds, shopped for groceries, matched slip-
> cover material, ate peanut butter sandwiches with her children, chauffeured Cub
> Scouts and Brownies, lay beside her husband at night. (p. 15)

In this excerpt, Freidan suggests that those who shop for groceries, make brownies, and drive the family carpool have a considerable stake in feminist activism as
well. Though her research emerges from conversations with women at Smith College, a somewhat exclusive institution, Freidan still teaches that advocacy for
women and women's issues can reside powerfully in the suburbs and everyday
lives of women from all sorts of backgrounds, not just in the academy among
feminist scholars. Freidan pushes readers to understand that, even if women
had not read leading feminist scholarship, they still would have been influenced
by the emergence of women's voices both in and outside of the academy. As
Freidan relates, women of the 1960s and 1970s just *felt* their need for agency—

for a voice in the ongoing social reform. Such a feeling can only be further accentuated today, as women take on more and more leadership positions in the academy. In fact, the number of women returning today to higher education continues to climb, so many of the women who baked brownies and drove the carpool in the 1960s and 1970s may emerge as tomorrow's leaders in the computers and writing community.

WHAT ABOUT THE INFLUENCE OF MINORITIES?

Last, cyborg history compels us to ask why minorities are not more prominent in accounts of the computers and writing community's history. Technological innovators like Evelyn Boyd Granville and Skip Ellis made prominent contributions to the development of computers and other technologies. Additionally, a number of minority figures played prominent roles in the social and cultural transformation of the 1960s and 1970s, and the substance and nature of their influence contributes meaningfully to the computers and writing community as well. Calls for new constructions of race emerged from seemingly very different forces, including the developing political agency of key leaders like Martin Luther King, Jr. and of resistance groups like the Black Panthers. If we include the contributions of minorities more prominently in histories of computers and writing, then we have more careful and responsible histories.

Minorities contributed actively as innovators in science and industry during the 1960s and 1970s, important efforts too often forgotten in histories of the computers and writing community. Evelyn Boyd Granville, just cited for her contributions as a woman to technology development, also serves as a prominent African American in the sciences. In a 1989 *Sage* article, she writes about her early experiences and her choice to become a scientist:

> Fortunately for me as I was growing up, I never heard the theory that females aren't equipped mentally to succeed in mathematics, and my generation did not hear terms such as "permanent underclass," "disadvantaged" and "underprivileged." Our parents and teachers preached over and over again that education is the vehicle to a productive life, and through diligent study and application we could succeed at whatever we attempted to do. As a child growing up in the thirties in Washington, DC, I was aware that segregation placed many limitations on Negroes. (We were not referred to as Blacks in those days.) However, daily one came in contact with Negroes who had made a place for themselves in society; we heard about and read about individuals whose achievements were contributing to the good of all people. These individuals, men and women, served as our role models; we looked up to them and we set our goals to be like them. We accepted education as the means to rise above the limitations that a prejudiced society endeavored to place upon us. (p. 44)

Granville's experience matches that of other minority innovators, all of them pursuing their dreams despite the challenging social and cultural oppression around them. Skip Ellis, who currently is on the faculty of the University of Colorado, was the first African American to earn a doctoral degree in computer science, earning his from the University of Illinois in 1969. After graduation, Ellis worked with supercomputers at Bell Laboratories, continuing work that had been the subject of his dissertation. Another important minority innovator doing work with supercomputers in the 1960s and 1970s was Roscoe Giles, who was an assistant professor at MIT from 1977 to 1983 and now is on the faculty at Boston University. Giles's research in the late 1970s emphasized the supercomputer's implications for physics and materials science. Last I would note that another African-American pioneer, Mark Smith, earned the first PhD in Computer Science from MIT, completing his studies in 1977.

Computers and writing histories should address the way large-scale change of any nature shapes everyone's lives, not just those of a few, a lesson 1960s and 1970s minority rights activists teach especially well. A leader often associated with protest movements for African-American and minority rights is Martin Luther King, Jr. Perhaps best known for his "I Have a Dream" speech, given on August 28, 1963, in Washington, DC, his vision includes much more, and he worked publicly and powerfully to fight for racial equality in the civil rights era, work that profoundly influenced individuals in that era and that continues to be influential today. In *Why We Can't Wait,* King (1964) outlines his sense of how social reform became such a powerful force in the 1960s and 1970s:

> Just as lightning makes no sound until it strikes, the Negro Revolution generated quietly. But when it struck, the revealing flash of its power and the impact of its sincerity and fervor displayed a force of frightening intensity. Three hundred years of humiliation, abuse and deprivation cannot be expected to find voice in a whisper. The storm clouds did not release a "gentle rain from heaven," but a whirlwind, which has not yet spent its force or attained its full momentum. (p. 16)

Among the most important aspects of this excerpt is that King, a man who publicly advocated for peaceful reform, articulated the storm of emotions that added an edge to social reform during his era. When individuals think about social change, they often find it convenient to view it from afar, treating it as a subject to be studied, not as an emotional experience, or they tend to localize any conflicts, deciding that those events with the most edge in local communities cannot represent the national or international scene and subsequently rationalizing that those most emotional national or international events do not have the same impact in their own communities. And this disengagement is a view King disallowed by showing that, although social change can only be understood fully by those involved, it affects and teaches everyone, a lesson especially powerful for computers and writing scholars today who do not always understand why many individuals in the world fear the changes that computers and other advanced technologies bring. In *Why We Can't Wait,* King also responded to the development and employment of the Atom bomb, offering a very specific reaction to the application of warfare technologies, the same sort of issues Marcuse raised in his flight from Germany, but more carefully articulated for an American readership in a time of social and cultural transformation. King's understanding of and advocacy for social revolution showed that sometimes those who are oppressed can assert sufficient agency not just to change their own life conditions, but also to educate and change the oppressors who must eventually come to terms with the diversity and inclusiveness in their states. King's voice and passion, then, necessarily influenced both majority and minority individuals in the 1960s and 1970s, helping them appreciate the complexities of change at that time, and King's perspective simultaneously helps individuals today to respect resistance and listen carefully to those both for and against technological progress. The best path for change, King taught, emerges in such a dialogic way.

Histories of the computers and writing community should draw from 1960s and 1970s activism to include a sophisticated understanding of how everyone is influenced and even implicated by racial politics, as such an understanding proves fundamental to discussions of access and diversity in the community. Perhaps no political organization had more agency in the move toward increased recognition of minority voices in the 1960s and 1970s than the Black Panthers, and their influence continues today. Two leaders of the Panthers, Kwame Ture and Charles V. Hamilton, wrote influentially about the proliferation of racism in the American society of that era and the need for reform. In *Black Power* (1967), they describe, "Racism is both overt and covert. It takes two, closely related forms: individual whites acting against individual blacks, and acts by the total white community against the black community. We call these individual racism and institutional racism" (p. 4). In articulating their sense of the dynamics of racism, Ture and Hamilton show both how to recognize it and how it is perpetuated; this sense of racist social structures gives a critical and careful voice to what many individuals had felt, but not known how to articulate and describe. It likewise helps individuals today to think about the way racism influences computers and writing: what form it takes, what influence it has, and how it can be engaged. Ture and Hamilton (1967) continue by characterizing the politics of social reform and resistance:

> When one forcefully changes the racist system, one cannot, at the same time, expect that system to reward him or even treat him comfortably. Political leadership which pacifies and stifles its voice and then rationalizes this on the grounds of gaining "something for my people" is, at bottom, gaining only meaningless, token rewards that an affluent society is perfectly willing to give. (p. 15)

In this excerpt, Ture and Hamilton suggest that racism and the process of reform are shared between those in the majority and those in the minority, and this observation establishes an important sense of the responsibility of all Americans in the civil rights era, responsibility individuals from around the world bear in their individual contexts today as well. Those in the majority cannot simply continue the status quo and do so without accountability, and likewise, minorities cannot expect reform just to happen. Instead, they must actively seek those victories for which they would call, both groups understanding that conflicts will evolve and need to be negotiated reasonably. The parallel for the computers and writing community is obvious: Those for widespread adoption of technologies and those against it must begin conversations and try to etch a middle ground fair to all or to articulate a reasonable rationale for agreeing to disagree. Ture and Hamilton's thoughts, like those of King, establish the continued applicability of 1960s and 1970s social reform for contemporary computers and writing. No person in the civil rights era could be simply an observer, a telling lesson for us today as we negotiate the increasingly technological world before us.

CONCLUSION

With the four questions discussed in this chapter, the concept of cyborg history has demonstrated that we should think more often about the 1960s and 1970s in histories of the computers and writing community. At the same time, it has shown us that we should ask questions of whatever histories prove prominent in the future, enabling a continual conversation about where we've been as it relates to where we are at that time as well as where we may be in the future. Having such a fluid and dialogic sense of the community's history should serve us well.

Community Voices

Christy Desmet

How did you come to be active in the computers and writing community?

It was a long and indirect path that brought me into the computers and writing community. In a way, I had belonged for a long time and just did not know it. When I sat down recently to read *Computers and the Teaching of Writing in American Higher Education, 1979–1994: A History,* I encountered people I had known for years: Lisa Gerrard, Michael Cohen, my old office mate, Carolyn Handa. I was a graduate student at UCLA during the late '70s and '80s, when the UCLA Writing Programs was just getting started. I worked with Dick Lanham, who was himself beginning to think about the issues that later would inform *The Electronic Word.* We tried out programs such as HOMER and read books about the potential and limits of computers. People in Writing Programs were also doing research on the students' use of computers, counting keystrokes and doing protocol analyses. I was not much involved with the technological end of Writing Programs, but I did write my dissertation with a clunky mainframe program that formatted dissertations automatically to UCLA's specifications. What a trial! We applied for internal grants to pay for time on the mainframe, then worked until all hours of the night because that was when computer time was cheap and terminals were available. During the 1984 Olympics, we had to get past an armed guard to pick up our output. But most of the time, my trials and tribulations were more mundane. Once, for instance, I saved a 90-page chapter over a 30-page one and spent July 4 retyping the lost text. Another time I waited one-and-a-half hours to tell a consultant that the title page could not print book titles in italics. She dutifully worked out a little program to fix that glitch; everything worked fine until I switched from a line printer to the laser printer, and then the entire title page printed out in Greek letters. The final reward for enduring this comedy of errors, however, was a laser-printed dissertation that cost only $40; I printed out the final copy of mine at 10:30 PM on my thirtieth birthday. When time came to submit the final document to the exacting officials of the Graduate School, however, it turned out that the mainframe program would not reproduce a copyright symbol on my title page; I was sent off in disgrace to find a typewriter that could do the trick.

Nevertheless, the nascent computers and writing community at UCLA must have had a real effect on my pedagogy. When I came to the University of Georgia in 1984, I had not lost all enthusiasm for technology, but I did not know where to get resources to use computers in my writing classes. I was not adept with a screwdriver and knew absolutely nothing about computer hardware. In my second year, I got some money from the Lilly Foundation to develop a Freshman writing class centered around the congruence between visual and verbal communication. I produced slides and transparencies and purchased a few films. I also bought fifteen copies of what had become the HBJ Writer, even though I had no earthly idea where my students could use this software. By the time we had a computer lab for the writing program, it was too late; the HBJ Writer was, by then, obsolete, and students marched dutifully into the labs to speed-type essays on a commercial word processor. Then my university became enamored of technology, providing an influx of money for hardware and software, but also institutional support of a more human kind. The Office of Instructional Support and Development at UGA, which was developed by folks I knew well from the Lilly Teaching Fellows Program, gave seminars on everything from web page creation to calculating grades in WebCT. With another small grant, I produced a lavishly colorful, if simple, WebCT site for the "Beowulf to Milton" course. Once again, the ability to use color and reproduce images lured me back to technology. At the 1998 Computers and Writing Conference in Gainesville, I gave a presentation about the pedagogy of using this Web site, and in particular, the ways students misunderstood its purpose. There, Carolyn Handa and I were reunited over margaritas and I was hooked, once again, on computers and writing.

Danielle Nicole DeVoss

Why do you choose to be active in the computers and writing community?

A year ago, a colleague of mine shared a story of her third-grade son, who, she was startled to find, believed that copying and pasting text and images from Web pages into a word-processing document was "writing." Four months ago, I worked with a junior-high student who constantly toted his laptop and required uninterrupted Internet access because he was working on projects at a distance for the Web-design company for which he worked. Today I read a quote from a literacy interview with a 15-year-old girl. After she provided an extensive list of the ways she uses computers in her daily life, she noted that "there is little use of dynamic technology in school, but I'm certain that this will change as the years progress."

These are not random examples, rather experiences like these are becoming more and more prevalent in my daily life. And these experiences energize, excite, and challenge me. Students come to the classes I teach with multiple and complex reading and writing practices, and dynamic expectations. I find this to be the most exciting aspect of teaching composition today, where all writing is computer-mediated, and more and more writing expands from text into the realms of media-rich compositions.

Without a community to support and sustain pedagogical experiments and new research approaches, however, scholars—no matter how dynamic—are sure to burn out, especially where ever-evolving technology is concerned. The computers and writing community provides colleagues, mentors, role models, and friends—thoughtful, critical, and dedicated individuals with stellar commitments to writing, to composition, to research, to computers, and to students. I feel at home here, and, as a member of this community, feel an obligation to continue and extend our collective scholarship.

Bradley Dilger

Why do you choose to be active in the computers and writing community?

I've always been encouraged by the accessibility of most of the people studying computers and composition who I've met at conferences, on mailing lists, or in online discussions like the Tuesday Cafe. In just a few years I have cultivated many relationships with researchers with similar interests. The result has been an eclectic mix of casual, experimental discussions—"Hey, what if we try THIS?"— as well as more methodical, formalized investigations that culminate in publication or conference presentations.

The genuine interest in collaboration and scholarly work of many people in the Computers and Writing community has benefited me greatly. While I've never had a problem getting support from my colleagues at the University of Florida, the strength of the wider community has also been a valuable source of encouragement.

Christa Ehmann

What's the most important aspect of the computers and writing community for you, and why is it so important?

As an educational researcher intrigued by the convergence of teaching, learning, and technology, I am most interested in a fundamental question: How, if in any way at all, can the Internet be used to enhance education? The computers and writing community provides me with focused opportunities to explore this issue. Studying various manifestations of online writing instruction helps me to develop systems for researching online learning and teaching as a whole.

The momentum of the online learning movement compels an examination of how core pedagogical principles can be applied to online contexts within and across disciplines. Although there are philosophical as well as practical challenges, my experience suggests that the Internet can indeed help to: (a) redesign writing environments; (b) diversify approaches to student and teacher development; and (c) engage writers and teachers in action research cycles that promote collaborative growth on individual and programmatic levels. Daily immersion in online learning also continues to transform personal assumptions about the distinctive nature of online learning and its relationship to other learning modalities. To that end, I appreciate the ways that the computers and writing community has highlighted the growing need for theory generating research about online learning that informs pedagogy in rigorous ways.

Patricia Ericsson

Why do you choose to be active in the computers and writing community?

For me, being active in the computers and writing community means listening intently to the conversations of this community through email lists, face-to-face conversations, conferences, and publications; it means contributing to the community via the same venues. I see Anthony Giddens' "duality of structure" exemplified in this community: the community helps to construct its members as the members work to construct it. I have found this community open to new ideas and willing to weave those new ideas with the old (which are never really old!) to create a rich tapestry of scholarship, lore, personal relationships, and more.

Choosing to be active in a community like this is challenging, exciting, and rewarding. It is challenging to encounter new ideas, fresh metaphors, that invite the reframing of questions. Those questions provide refocused lenses with which we can "see" things differently. A community that offers such challenges is exciting because it keeps scholarly interests alive and fresh, continually requiring its scholars to rethink and reconsider. The rewards of being active in this community are innumerable; however a few deserve emphasis. It is rewarding to work in a scholarly community that is not staid and certain. Such a community is welcoming to new scholars with novel ideas and approaches. Although the computers and writing community is growing, it is still small and congenial enough that it retains the bonds of a small, close-knit community. And finally, it is rewarding to work in a scholarly academic community that values teaching, particularly the teaching of communication skills.

I choose to be active in the computers and writing community because it affords me opportunities that I see as almost unlimited. This community continues to provide me with both the academic challenges and the personal rewards I need to stay interested and active.

Douglas Eyman

What scholarly project in computers and writing has been most influential for you, and why has it been so influential?

The computers and writing project that has been most influential to me, and which I have been privileged to be involved with, is *Kairos: A Journal of Rhetoric, Technology, and Pedagogy*. In 1996, I attended my first Computers and Writing Conference; there I met (among many other strange and wonderful people destined to be my friends and colleagues) a group of graduate students from a variety of institutions who had gathered together to enact a vision they shared—the creation of an online, peer-reviewed scholarly journal that would encourage authors to not merely theorize about hypertext, but to actually write it. This, to me, was exciting news, and I wanted to be a part of it. In the first issue (released shortly after the conference), I contributed a news article and a hypertext review of George Landow's *Writing at the Edge: Student Webs From Brown University*. In the second issue, I published my very first peer-reviewed scholarly article, "Hypertext And/As Collaboration in the Computer-Facilitated Writing Classroom." I joined the staff as CoverWeb Editor for issue 2.1; by issue 5.1, I had become chief co-editor (with James Inman). But the reason I identify *Kairos* as the scholarly project most influential to me is not only because it has played such a pivotal role in my professional development: *Kairos* has also granted me the opportunity to learn about a diverse range of issues and theories as they are being actively developed and debated in the field of Computers and Writing, from Queer Theory and Gender Issues to concerns about tenure and promotion; from Visual Rhetoric and Hypertext Theory to practical classroom applications of new media; from OWLs to MOOs—and all utilizing the medium in rich, exciting, and sometimes surprising ways. *Kairos* is also a testament to the rich sense of community that informs Computers and Writing; we have been able to provide excellent scholarship only because of the time and talent freely given by our editorial board and staff: *Kairos* is now entering its seventh year with no monetary support, and no individual institutional affiliation. I certainly hope that it will continue to thrive, and continue to present new and exciting scholarship to the field of Computers and Writing.

Dawn Formo
and Kimberly
Robinson Neary

Why do you choose to be active in the computers and writing community?

Dawn: My scholarship in computers and writing started in very practical ways. As the co-editor of *The Writing Instructor,* I helped transform the journal from a print journal to a digital one. In the fall of 1999, faculty on my home campus received a state initiated grant to collaborate with the faculty and students at six neighboring high schools: the Collaborative Academic Preparation Initiative (CAPI). The goal of the grant is to reduce the need for remediation in math and English among incoming university students. With experience as a former writing center director and with the desire to reach out to as many college-prep (non-AP) high school students as possible, I designed a CAPI website and my campus' first OWL.

With the launch of the OWL, I asked myself, "How would we teach OWL consultants and high school student users of the OWL to engage in lively discussions about their OWL submission?" This question has led not only to the development of an OWL pedagogy for consultants and students alike, it has also piqued my research interests in gender and discourse development and analysis.

Kimberly: I was a graduate student doing work in composition and gender. One of my early research projects involved observing classroom interactions and analyzing girls' papers to trace the effects of gender on the writing process. After conducting a small case study, I knew the question of gender was significant, and I wanted to continue my research on a larger scale. Fortunately, Dawn's work with CAPI gave me a research database to conduct my work. The unique nature of this database also begged for another level of analysis.

Dawn and Kimberly: With our shared interests, together we wondered, "What happens when the OWL environment meets the techno gap?" In *Gender Gaps: Where Schools Still Fail Our Children,* the American Association of University Women reports: girls seem alienated from technology for a range of reasons including access and gender expectations. We want to find ways for girls to maneuver such barriers. Even more, we want to learn from the girls. Might there be ways that technology allows and encourages girls to strengthen their rhetorical stance in digital spaces?

82

Yet, while OWLs offer many potential benefits for all students, they could become a space in which girls, in particular, are silenced. As we look at who has access and who is encouraged to use computers, we wonder if the online environment has the potential to inhibit the writing process becoming, in Audre Lorde's terms, "the master's house."

We participate in the computers and writing community because we are both terrified and exhilarated by the community's potential to explore these divergent paths. In collaboration, we share the fear and joy. Together, we want to be connected to scholars who theorize and practice the future. We want to work in concert with others to make sure that gender is part of these discussions, constantly keeping in mind that the different socialization processes of males and females will most likely result in different experiences both in the writing classroom and in the electronic environment.

Lisa Gerrard

How did you come to be active in the computers and writing community?

In 1980, when I started teaching with computers, there wasn't a computers and writing community, at least not at the college level. I knew about Hugh Burns' dissertation, Ellen Nold's article, and several software development projects for elementary school students, but I was pretty isolated with my classroom experiments—requiring students to use a mainframe computer and text editor for writing. I discovered that other college instructors were exploring computers and writing in 1982, when I began working on WANDAH, writing software (a combination word processor-prewriting program-revising program) a group of us developed at UCLA. A number of English instructors from other schools came to UCLA to see WANDAH, giving me my first inkling that the seeds of a community were out there. That year, Lilly Bridwell and Donald Ross invited members of our development team to talk about WANDAH at what became the first Computers and Writing Conference—a small (about 10 people) meeting of software developers held at the University of Minnesota in October, 1982.

A year and a half later, in April, 1984, I attended the sequel to this conference, again at the University of Minnesota. And it was a wonderful meeting—the first conference I'd been to where everyone was excited about learning. I didn't notice the careerism and self-promotion that I was accustomed to seeing at academic meetings; instead the participants were eager to find out what others were up to and to share their successes and failures. We were all there, all 250 of us, to connect with others who were trying similar experiments, largely alone on our home campuses: growing our own software, studying how computers affected student writing, deciding how to spend grant money, wondering how to connect a lab to a writing course. We were all using different (and, of course, incompatible) hardware and software, we were learning from one another, and we were becoming a community. A very exciting time.

The two conferences had been requirements of Lilly and Don's FIPSE grant, and were not intended as the start of a series. Towards the end of the conference, I blurted out to Lilly that we should have a conference like this every year. By which I meant *she* should have it every year. But Lilly called my bluff: "Good idea, why don't you organize it next year?" So I did. And people came enthusiastically. Like me, many of them were the oddballs of their home departments and needed a forum where they could share their experiments and be, well, less odd. In the first few years of computers and writing, it was the conference that provided this forum and that made us a community.

Paula Gillespie

How did you come to be active in the computers and writing community?

For a long time the electronic community was my only professional community. As a writing center director in a medium-sized school, I had no colleagues in my field to chat with over coffee, no one to sound ideas off, no one to ask for advice. I signed onto WCenter and found a vibrant community that has been not only a source of professional mentoring and counseling for me, but that has furnished friends, co-authors, co-presenters, and abundant good ideas. Contributors to WCenter have read articles with me, have challenged my canonical thinking about things, have suggested research directions, have discussed, debated, and challenged writing center theories.

A few years ago, a CIWIC workshop at Michigan Tech took that engagement in computers and writing a couple of quantum leaps forward, making new technologies accessible, available, and downright feasible. Not to mention fun. And on reflection, I think it's this element of fun that makes all our learning breakthroughs exciting. Now more than ever my community includes my tutors and students, on discussion boards and in chat rooms for virtual class meetings.

Morgan Gresham

What scholarly project in computers and writing has been most influential for you, and why has it been so influential?

There have been two recent projects that have greatly influenced my thinking as a computers and composition scholar. The first is Patricia Sullivan and James Porter's *Opening Spaces: Writing Technologies and Critical Research Practices. Opening Spaces* helped me articulate, for my dissertation, my understanding of computers and composition as landscape. Without their influence, I would not have been able to negotiate a sound theoretical grounding, a base, a key to understanding the map, only to be lost in this web, this landscape that is a languagescape constructed from the interactions of private and public, personal and political. If we continue to think in terms of identifying and classifying political systems into dichotomies we will limit the potential of computers to redefine those structures. The confluence of post-modern mapping theory and feminism helped me refine what it means to be a feminist researcher in composition. For example, when Sullivan and Porter argue for the situatedness of research, suggesting that we must work to insure that research methods "don't drown out alternative voices," I knew I could no longer be content dismissing students voices from my work, especially when it was their concerns about technology that were deeply influencing the shape of the research.

The second is Kristine Blair and Pamela Takayoshi's *Feminist Cyberscapes: Mapping Gendered Academic Spaces.* It brings together multiple perspectives on feminism, computers and composition without a pre-determined view of feminism. Because feminism takes many definitions in *Feminist Cyberscapes,* I get to see myself as a feminist student, researcher, scholar, teacher, and computer user. Also, in their introduction Blair and Takayoshi establish the importance of reconsidering the binaries through which we have often studied computer and composition.

Overall, in these two texts we see the mainstays of computers and composition studies: the complexities of more critical, qualitative studies and the textures and personalities of qualitative discourse analysis.

Dene Grigar

How did you come to be active in the computers and writing community?

What drew me into the Computers and Writing Community was the prospect of "community," the idea that being a scholar in this field would provide me with an extended group of family and/or friends that offered guidance, support, and leadership for our work and career goals. But what kept me involved in the CWC was the fact that the CWC made little distinction between graduate students and faculty, and between different types of faculty (tenured, adjunct, non-tenure-track) and that people are paid respect whether they publish in print extensively or are new to the field. It was these qualities of the CWC that I experienced firsthand at my first Computers and Writing Conference in 1993. At that conference I watched Michael Joyce and graduate students he was working with give a wonderful presentation on hypertext together. As a graduate student myself I was intrigued by the idea that faculty would collaborate with students without drawing attention to the difference in their academic levels. I was later amazed at the kind of assistance and support for my own ideas I received from members of the audience after my own presentation at that conference: Jeff Galin and Joanie Latchaw invited me to submit my work for their collection, *The Dialogic Classroom.*

As I joined the listservs associated with the CWC, such as MBU and later ACW, I found the most seductive aspect of the CWC, however, were the friendships, partnerships, and collaborations it made possible. No matter how alone I might have felt in my graduate program or in my faculty position at Texas Woman's University, I knew all I had to do was post a note on the listserv and I would find many friends to help me sort out problems with my classes, work through an idea for an article, or work through strategies to get through a tough project.

Sibylle Gruber

Why do you choose to be active in the computers and writing community?

A while back, I read a piece by Keng Chua who pointed out that women and minorities might find ourselves "being shaped and managed once again through the new technologies," where we have to follow a prescribed online identity and fixed online discourse conventions. This statement made me realize that we might get too complacent about our technological achievements, that we might not look as closely and as carefully at some of the online practices that perpetuate typical and stereotypical behaviors—machismo, capitalism, colonialism, aggression, violence, oppression, authoritarianism, and so on. Working closely with many members of the computers and writing community has made me realize that many of us are concerned with these issues, and that we can work together to address the problems that are part of integrating new technologies into our theories and practices.

Much of my research on technology is influenced by feminist and cultural theories, working toward a better understanding of women's cyborg lives, the connections between online and offline identities, issues of power and access, as well as online body politics. I have been fortunate to be introduced to these issues by Gail Hawisher and Cynthia Selfe early on in my academic career. I have also been fortunate to work closely with many of the feminist scholars in the computers and writing community, engaging in listserv discussions, working with them on edited collections, and being re-energized by their words and actions at conferences. It is a community that allows for constant development of thought, and that encourages me to continue my scholarly endeavors into women's issues online.

Christina Haas

How did you come to be active in the computers and writing community?

I came to be involved in the community of computers and writing in much the same way that I came to understand why studying technology is itself important. In both cases, it happened something like this: a nice person saw me wandering around, understanding little, and gently nudged me in the direction I needed to go.

I have often told the story about how Richard Enos, in a seminar on Classical Rhetoric at Carnegie Mellon gently opened my eyes to the profundity of the questions of materiality that underlie all research on technology. The simple, yet profound question that Rich helped me identify was this: what happens to human thinking and human culture when the material means by which we communicate change? I was all wrapped up in word processing and crude early email at the time, but one day in class Rich put into my open palm a clay tablet and a stylus—and said something like "Imagine writing your dissertation in this medium," and the profundity of what technological change means for writers struck me.

A year or two later, still a grad student, I had given a presentation on some of my early research at one of the early Computers and Writing conferences. Feeling irrelevant, misunderstood, out of place, poorly dressed—all those things grad students (and everybody else) feel at conferences. When out of the crowd comes this elegant, smiling woman who seems to be parting the crowd and seeking me out. She takes my arm, looks directly into my face, and says, "The kind of work you are doing is exactly what this field needs. Now, how would you like help me to put together a panel?" One minute I was a self-absorbed nobody and the next, I was part of Gail Hawisher's next big thing. Another strong and gentle soul, leading me into a place I hadn't known was there, but—it turned out—was exactly where I needed to be.

Marcia Peoples Halio

How did you come to be active in the computers and writing community?

I became active in the computers and writing community in the early '90s because I was concerned that the machines were having an adverse effect on students' writing (and our own). I thought we were spending too much time formatting and decorating our masterpieces rather than thinking about the content. Quickly, the community asked me questions I needed to hear and helped me to think more clearly about the effects of the machines on the messages we produce. In the last ten years, I have benefited immensely from being a part of and lurking in countless discussions about computers and writing, and I have seen first-hand the power of networks to link questing souls. Without the community, I would still be howling in the wilderness!

Carolyn Handa

How did you come to be active in the computers and writing community?

My involvement with the computers and writing community was a fortuitous accident. Because of a graduate class in concordance compiling that I enrolled in while finishing my degree at UCLA, I happened to do some work with computers in the days of the good old mainframes. That involvement, however, led me to teach writing in a few experimental writing classes using mainframes when I began my first teaching job as a new PhD at the University of California, Davis. To make this part of the story short, I eventually became one of a group planning and teaching in the first computer classroom at UCD. I was then one of three instructors who spoke about our classroom at a 4Cs conference in Atlanta in the late '80s.

At the Cs, I attended every computer presentation I could find. I heard Cindy Selfe, Tommy Barker, Hugh Burns, Trent Batson, Diane Langston, Carolyn Boiarski, and Kathleen Skubikowski, among others. I got the idea for my book *Computers and Community: Teaching Composition in the Twenty-first Century* at this conference. Once I worked with all the good people who contributed essays to my book, I was hooked on the computers and writing community. I had never before met such a group of generous, enthusiastic, supportive people. And I have stayed active in this community over the years because of the generous spirit that I first experienced as a newcomer. Thank goodness for happy accidents.

Linda Hanson

Why do you choose to be active in the computers and writing community?

At the heart of the work we do in composition studies resides the belief that we can continue to advance literacy. The optimism to act on that belief, however, is most apparent in the computers and writing community. Since the earliest meetings at CCCC's, folks in the computers and writing community have yoked emerging—and still rapidly changing—communication technologies to pedagogical tasks, thereby changing both. Our respect for one another's work, for the diverse approaches to implementing, assessing, and representing our work, uniquely marks this community of scholars. What brought me into computers and writing, what keeps me coming back, is the vitality of a reflective, sharing, intellectual community grounded in the pragmatic concerns of teaching and learning in our own classrooms and the larger community.

Muriel Harris

How did you come to be active in the computers and writing community?

When I think about how and why I came to be active in the computers and writing community, I realize that as computers became an integral part of writing and writing instruction, my involvement was a natural extension of working with writers in our Writing Lab at Purdue University. Looking back, I also see that my experience reflects some of the ways that computers entered into tutorial instruction. I was drawn into this world long ago in the Dark Ages of Technology, B.D.W. (Before the Dawn of the Web). At that time of primitive electronic interaction, a graduate student tutor and I worked with writing sent to our Lab via modem from a class in one of our Schools of Engineering. We ran papers through a readability software program, "Writer's Workbench" (developed at Bell Labs to assist technical writers to revise effectively) and then sent back tutorial questions and advice to each writer based on the results of the readability scale. When e-mail became available, it seemed an appropriate outreach medium for Purdue students to contact us during the hours our Lab was closed, at a time when they were also more likely to be writing their papers (e.g., late Sunday evenings). We weren't able to reply immediately, but we did try to respond within twenty-four hours. As it became evident that late-night writers also needed access to our cabinets filled with instructional handouts to accompany tutorials (various rules of grammar, MLA and APA formats, advice on rhetorical matters, etc.), we put the handouts online in ASCII character format, with an instant auto-respond program so that a request for a handout was answered within a few seconds. This was Version 1.0.0 of our OWL (Online Writing Lab). We hadn't anticipated the world-wide interest in this service because we weren't fully aware of the global hunger for instructional materials on writing in English. With the advent of Gopher, we created an OWL Gopher site, with links to useful resources that were sprouting up. And I should acknowledge that the rapidly increasing usage of our e-mail and Gopher service was primarily from off-campus. Purdue students were apparently not yet adept in using e-mail, not comfortable with it, or not accustomed to it as a medium for writing assistance. Other writing labs experimenting with the world of computers and writing were finding similar results. But we did add a few computers to our Writing Lab tables for the students who were beginning to bring in drafts of papers on disk. And thus as computers began entering into how students were writing, computers also came into our Lab tutorials, along with our OWL site.

Then a very computer literate graduate student (for whom I managed to find funding in the face of great administrative skepticism about why computers should be involved with writing) introduced me to the early stages of the World Wide Web. Using Mosaic (the only browser available), we built an OWL Web site (http://owl.english.purdue.edu) based on what we had developed in earlier formats. We reformatted the dozens of instructional handouts that had been on Gopher, created pages of links to online resources, gathered a list of search engines for easy access, and experimented with different ways to approach tutorial services online. Not finding effective ways to incorporate what we considered essential components of face-to-face tutoring, our OWL was still primarily an additional resource to accompany tutoring and a tool for outreach services. However, I was inevitably being drawn deeper into the world of computers as computer technology wove itself into other aspects of our Lab's work, both in terms of what we could offer student writers and what we were learning from developing our online presence. In succeeding years, our OWL has developed as new technology offers us new opportunities to integrate computers into our expanding repertoire of tutorial services. Being part of the computers and writing community now means (to me, at least) interweaving all facets of writing assistance that we can offer students, instructors, and Internet users. It also means a fascinating new area for those of us in writing labs in which to engage in knowledge-making about writing centers' use of technology. I can no longer conceive of how we can think about writing or tutorial interaction with writers without including computers as part of the tool box with which we work.

Bill Hart-Davidson

What scholarly project in computers and writing has been most influential for you, and why has it been so influential?

While I have found inspiration in many specific projects in computers and writing, one that has really made an important difference in the direction of my own work is Hugh Burns and Paul Leblanc's *Writing Teachers Writing Software: Creating Our Place in the Electronic Age.* This book offers case studies of people putting their expertise and experience as writing instructors to work building software to support writing instruction. Reading the book in 1993, just as I was teaching my first writing courses in a networked computer lab and not long after getting my first glimpse at an odd program called Mosaic, I saw a career's worth of opportunity. At that moment, for myself and my students, writing on computers still meant composing print-format texts with *Word 4.0* or *MacWrite,* or maybe posting to a UseNet newsgroup. Still, as a result of Burns and Leblanc's work, I began to notice that the tools and environments in which I was writing, and those in which I was encouraging my students to write, were built out of the very stuff that I was supposed to be in command of as the teacher: assumptions, theories, principles, and pedagogies of writing.

Writing teachers writing software? Of course! It's as easy a fit as Engineers writing CAD packages, right? But how much higher the stakes? Writing software—that is, applications which support invention, arrangement, style, memory and delivery of written discourse—was, even in 1993, arguably the most often used software. E-mail alone . . . well you get the idea.

Fast forward a bit to the end of 1998, the awakening of the World Wide Web in American popular culture. More people than ever doing "computers and writing" or "writing on computers," however we might call it. And more writing courses than ever conducted in computer facilities. I re-read *Writing Teachers Writing Software* as part of my dissertation research, which wrestled with the question of how best to prepare teachers to work in networked spaces. I titled the concluding chapter of my dissertation "Writing Teachers, Writing Networks," echoing Burns & Leblanc and attempting to capture the enormity of the opportunity I believed, and still believe, the computers and writing community faces. My study, which used methods from participatory design to develop a training program for first-time teachers of professional writing in network environments, gave me the opportunity to articulate this opportunity in the "conclusion" to the dissertation:

Teachers of writing, it might be argued, are responsible for the "core technology" of writing networks—writing itself. This realization is not a conclusion, per se, but is rather an invitation to begin exploring the connections between and across the disciplinary identities of writing specialists and computer specialists, systems and software designers.

The exploration I envisioned is what most accurately describes my current work in the field. But this line of inquiry didn't originate with me. I see myself continuing a project begun, years earlier, with *Writing Teachers Writing Software,* a project with the aim of "creating our place in the electronic age."

Gail E. Hawisher

What scholarly project in computers and writing has been most influential for you, and why has it been so influential?

Interestingly, each of the scholarly projects in which I've engaged these past 20 years, almost without exception, have underscored for me the importance of the computers and writing community. Each project, in fact, has drawn me more deeply into the community. Through my work on word processing, I met Lilly Bridwell-Bowles, Lisa Gerrard, Chris Haas, Pat Sullivan, and Cindy Selfe, all at our early Computers & Writing Conferences in Minnesota and Pittsburgh. My research on computer-mediated communication brought me in contact with Trent Batson, Fred Kemp, Paul LeBlanc, Michael Joyce, and Charlie Moran. The Austin Computers and Writing Conference facilitated those face-to-face meetings.

When Charlie, Cindy, Paul and I began work on our history of the field, we had even more opportunities to meet more of the community, and this time we were able in our book *Computers and the Teaching of Writing in American Higher Education, 1979–1994* to feature the new voices of the community, voices that today are old favorites—Eric Crump, Becky Rickly, Johndan Johnson-Eilola, Locke Carter, Michael Day, Pam Takayoshi. In that book, we tried to demonstrate the importance of the community by listing on the inside cover all those who had influenced the project.

And the list goes on. When Pat Sullivan and I began research on the *Woman@WayTooFast* project, the community again welcomed us. Sibylle Gruber, Marilyn Cooper, Kris Blair, Susan Romano, Carolyn Handa, and Elizabeth Sommers are only a very few of the women who helped us out on that project. And now, as Cindy and I become more involved in our research on how people acquire the literacies of technology, we've come to deeply appreciate the graduate students with whom we've worked over the years and who today are part of the computers and writing community. Mary Hocks, Sibylle Gruber, Tricia Webb, Mary Sheridan-Rabideau, Joyce Walker, Liz Rohan, and Karen Lunsford number those from Illinois. More recently, too, with *Computers and Composition* moving to an international venue and with these scholarly projects being read globally, we've come to appreciate our colleagues from all over the world. Chris Pringle, Dimitris Koutsogiannis, Bernie Susser, Ilana Snyder, along with other colleagues in Great Britain, Japan, Norway, Egypt, and Zimbabwe, have played important roles in our scholarly lives. Thus while I deeply value the research—the scholarly part of my life—it would have meant far less to me without the people. I can't imagine my life in the profession without them. Many thanks to James Inman, another great colleague, for the opportunity to write these words.

Byron Hawk

How did you come to be active in the computers and writing community?

I came to be a part of the computers and writing community quite by chance. In 1995, the year I started my PhD, the Web was just emerging as an important cultural and educational space. I happened to be in a program with other graduate students who were working with technology (such as Collin Brooke and David Rieder), so technology just became a regular aspect of intellectual investigation into teaching writing and the fields of rhetoric and cultural theory. I learned most of what I know about computers from other graduate students while working on various projects, often outside of the context of specific courses or coursework. In 1996, David Rieder and I decided to start an online journal that focused on culture, rhetoric, and new media. We got together, brainstormed a name and a to-do list, and sent out a CFP. Continued work on the journal (*Enculturation*) got me involved in issues of rhetoric and web design, argument in hypertext, electronic publishing, and visual rhetoric. As the WWW exploded to become a predominant cultural medium, these areas of study became more important and pushed my interest in pursuing digital writing and culture further. Participating in the Computers and Writing conference became a given, and it is really through Computers and Writing that I became a member of the larger community. The conference gave me the opportunity to see what other people were doing in the field and opened the way for me to begin collaborations with other online journal editors. The chance occurrences of becoming a part of the local community at the University of Texas at Arlington, working with Victor Vitanza and the other graduate students there, at a time when the Internet became a household entity, all set the stage for becoming a professor who works and teaches with new media and participates in the computers and writing community.

Cynthia Haynes

What worries you about the computers and writing community, and why does it worry you?

"Computers and Writing" is at a crossroads, and not one that necessitates a move this direction or that. We do not lack for interesting and challenging trajectories; we have even invented a few ourselves. No, the intersection in which we find ourselves is a bit more abstract. In fact, it is a conjunctive problem. I'm speaking about the "and" and its diminishing power to join "computers" with "writing" in any meaningful way, its merely nostalgic semantic function in linking our past to our future. 'And' taking the question beyond the worry that denotes concern, I think the 'and' needs 'worrying'; it needs to be disturbed, set in motion instead of standing guard between techné and language.

But setting the 'junctive' against its 'con' might disclose a lurking redundancy. And there's the rub. Redundant—meaning the same; without a rhetorical purpose, or worse, no longer needed. It could be the most significant question to which we turn because it would shift our gaze into the center of our own cocoons. Yet, what delicate meta'morph'osis awaits us? The answer is on the far side of the 'and' . . . and proximity is everything.

TyAnna Herrington

What's the most important aspect of the computers and writing community for you, and why is it so important?

The most important aspect of the computers and writing community for me is its attitude.

The Computers and Writing community embodies the same attitude that I find in a close friend. It is

Inclusive but scholarly, although not stuffy . . .

Intelligent, not arrogant . . .

Unconventional, innovative, creative . . .

Concerned, caring . . .

Loyal, supportive. . .

Rebellious but mindful and careful about its effect on others. . .

Generous . . . helpfully, hopefully, and unabashedly generous . . .

Diligent and undeterred—necessarily hardheaded but intellectually open . . .

Fun. Way fun. And most of all . . . joyful.

The computers and writing community has provided me space—intellectually, professionally, and personally—to think what I think, to teach what I teach, to achieve, to create, and to keep learning . . . and keep learning . . . and keep learning. In turn, it has also let me make all this available to my students, and will someday, I hope, to their students. Its attitude creates a circle . . . of friends, of colleagues, of students, of intelligence, scholarship and educational progress.

Brooke Hessler

What's the most important aspect of the computers and writing community for you, and why is it so important?

My perspective on teaching with technology changed dramatically when I moved into Web environments. I found myself viewing my classroom, my students, my pedagogy, my institution, and myself differently—all as texts on a screen. Beyond everything else we are, we become rhetorically charged artifacts that are co-constructed and re-constructed over time—often with unexpected consequences. In the computers and writing community I've found people who share my fascination with this phenomenon and my desire to do something with it, for it, and about it.

Beth L. Hewett

What scholarly project in computers and writing has been most influential for you, and why has it been so influential?

Thomas Barker and Fred Kemp's "Network Theory" has influenced my thinking a great deal. To me, their ideas about "textualizing" a class and the potential values of students reading each other's imperfect text (as opposed to their general experiences of reading schoolbook or professional text) link inherently with those of Robert Zoellner, who probably never imagined networked computers. Instead, Zoellner imagined the value of a "wrap around" blackboard that would enable students to write for and with each other. Just as important, however, he challenged teachers to write spontaneously for their students, letting them see the genuine processes of a real writer—warts and all. These ideas have helped me to become a more open thinker about how computers and writing may benefit students through real-time "imperfect" textual interactions with each other and with their teachers. Both asynchronous CMC and synchronous interactions, such of those of a whiteboard or other text-sharing platform, may enable the hopeful results that Barker and Kemp, as well as Zoellner, "theorized" some time ago.

My own research relies heavily upon my developing sense that students should have many opportunities to write to each other and to their writing teachers, or coaches, about their writing. Like Barker and Kemp and Zoellner, I hope that such talk, which implies both reading and writing imperfect text, will lead to more critical thinking, reading, and writing skills. Yet, these ideas, however common sense in nature, seem largely untested and unproven; how novice writers use CMC writing and talk to improve their communications is still unclear. I look to the computers and writing community to test these ideas in systematic ways that can help to unlock the secrets of writing development when CMC is used to practice and talk about writing.

Krista Homicz

What's the most important aspect of the computers and writing community for you, and why is it so important?

The computers and writing community profoundly shapes my professional development, especially at this formative time during my graduate school career. As my graduate program does not offer a specific concentration in technology and the use of computers in composition, the computers and writing community informs my graduate scholarship at a distance, providing me with resources, ideas, issues, and feedback that are essential for focusing my thinking and scholarship in this area. Although the members of this community are often located in universities that are disparate geographically, I often feel very close to these colleagues in terms of online collaboration. The community's online resources present many ideas that help and encourage me when developing strategies for integrating technology with teaching. Discussions and inquiries of scholars on the TechRhet email list-serv provide daily immersion in a discourse community and contact with these professionals and ideas to which I would not otherwise have constant and familiar access. Many of these scholars are leaders in envisioning applications for the newest technologies in teaching, and they are the first to share their emergent knowledge and expertise publicly via web pages and MOO discussions. The community's online resources have also been useful in my research. While working on projects associated with technology at my home institution, I have often been able to draw upon the work of computers and writing professionals that are involved in similar projects at their respective institutions, both via online discourse and through direct discussion at conferences. The intimacy of the group, in terms of size and supportiveness, is one of its greatest attributes, affording me with many opportunities to develop and refine ideas, discuss ideas of mutual interest, and to work jointly on projects.

Christine Hult

How did you come to be active in the computers and writing community?

For some obscure reason, my interest was captured by computers at the outset of my career, perhaps due to the torturous task of re-typing my dissertation a hundred times in pre-computer days. My first professional book, co-authored with Jeanette Harris, was titled *A Writer's Introduction to Word Processing,* published by Wadsworth in 1985, at the very inception of the computers and writing field that happily coincided with the start of my own post-graduate career. Ten years later, in 1995, with my scholarly interests still firmly centered around the intersection of computers and writing, I co-wrote a large grant proposal to design and offer a freshman English course delivered to students via the Internet. Of course, at that time, the Internet meant email, so, after receiving the grant, we designed a program that used email to teach those first online classes. As we evaluated our tentative steps toward online teaching, we were struck by a couple of intriguing factors: 1) Students with various disabilities (e.g. hearing-impaired students, blind students, or even those with a temporary disability like a broken limb) seemed to thrive in the online environment; 2) Students who otherwise would be locked out of access to higher education were able to take college classes from remote locations and on their own time.

The evaluation of online teaching and learning became my passion, with the ultimate goal of helping teachers to discover "best practices." Since those first very rudimentary online teaching efforts, our department and university's commitment to and understanding of online education has blossomed into a multi-faceted initiative to reach out to various under-served student populations. I find working with these diverse student groups, as well as working with a variety of faculty members in many disciplines, extremely satisfying. In particular, in our master's program in Technical Communication, I teach practicing technical writers from around the globe, many of whom are located in remote locations and who telecommute to their jobs, in addition to working on an advanced degree entirely online. I have found online teaching to be as intellectually rewarding and challenging as any teaching that I've ever done. I have also found the puzzle of how to best teach in this environment and how to evaluate the effectiveness of teaching and learning online to be a compelling scholarly enterprise.

Sue Hum

What scholarly project in computers and writing has been most influential for you, and why has it been so influential?

Computers complicate the teaching of literacy, both physically and intellectually, changing how we make meaning. Thus, technology cannot be treated simply as an ingredient that we add and stir into the soup of classroom curricula. Instead, technology must be considered a volatile catalyst that holds the potential to transfigure academic and culture experiences. In order to be more effective teachers of writing, we must also recognize how various electronic spaces, constituencies, and technologies affect literacy education.

Don Ihde's theories have been most influential in helping me explore the mediating influences of technology, in particular how technology is transformational and transfigurational, how it "teaches" ways of seeing and thinking. A professor of Philosophy, Ihde explains that technology is the "non-neutral means by which an entire cultural gestalt can be carried." Particularly enlightening is Ihde's analysis of how Galileo's vision is technologically embodied through a primitive telescope, providing a suitable analogy for considering how technology, as a medium, might mediate and influence the information it delivers. Ihde's work accounts for how technology influences behavior and perception, how technology inscribes and creates while functioning as a medium.

James A. Inman

What's the most important aspect of the computers and writing community for you, and why is it so important?

As I hope the community voices segments of this book make evident, the strongest aspect of the computers and writing community is its people. From graduate students and adjunct faculty, to technologists and professional writers in industry, and to full-time faculty and administrators, the community is fortunate to have many people doing exceptional work. All the people featured in this book took time out of their already too-busy lives to write responses, and they trusted me not just with those responses, but also with pictures of them; this trust well represents the generous spirit of the community.

At the same time a number of people responded, others were unable, and I want to be sure to recognize them here because they too play a vital role in computers and writing; their schedules simply would not allow them to prepare a response, or I somehow missed them in the various sources I was using to locate prospective respondents. Rather than name specific individuals—an enterprise that would always have me unintentionally leaving out as many names as I could intentionally include—I hope this serves as a statement of general appreciation for all that everyone has done. This community works because of you.

3

Recasting Canonized Computers and Writing Scholarship, 1979–2000: Cyborg Narratives

In the cyborg era, scholars in the computers and writing community often struggle for acceptance of their work in institutional and organizational contexts, and it has been this way since the beginning. These struggles have been documented and explored in any number of forums, from journals like *Computers and Composition, Kairos: A Journal of Rhetoric, Technology, and Pedagogy,* and the *Journal of Electronic Publishing,* to professional organizations like the Alliance for Computers and Writing (ACW) and the Conference on College Composition and Communication (CCCC). In general, the conditions have stayed constant: Those in the computers and writing community write often from the margins of their institutions as part-time faculty, nontenure-line faculty, or graduate students, and those more senior and/or in more powerful institutional positions often have limited respect for the depth, breadth, and rigor of these community members' work. Consider the following excerpt from Cynthia L. Selfe's keynote address at the 1998 convention of the CCCC, in which she talks about the way scholarly discussion about computers often draws uninterested reactions from those not invested in the computers and writing community:

> [Technology] . . . is . . . the single subject . . . best guaranteed to inspire glazed eyes and complete indifference in those portions of the CCCC membership, which do not immediately open their program books to scan alternative sessions or sink into snooze mode. . . . I can spot the speech acts that follow a turn of the conversation to computers—the slightly averted gaze, the quick glance at the watch, the panicky

look in the eyes when someone lapses into talk about picoprocessors, or gigabytes, or ethernets.

Put simply, time has proven that scholars outside of the computers and writing community may be unwilling to pay attention to its scholars' work, a problematic reality that means the struggle for acceptance will likely continue into the future.

This chapter suggests that, although scholars unwilling to learn about the computers and writing community and its work are part of the problem, we can also help ourselves by thinking carefully about scholarship already visible in the community, attempting to learn about ways such work can encourage and support present and future scholars with mutual interests. Doing so enables us to emphasize not just those who contribute work to multiple areas of the computers and writing community, but also to emphasize those who offer expertise and experience in a single area of two. In the present discussion, I first define *cyborg narrative*, this chapter's key term, by using postmodern and postindustrial perspectives on legitimation, and I then describe two cyborg narratives I believe can be sketched from visible community scholarship between 1979 and 2000. I specifically choose 1979 because that year has often been cited as the birth of computers and writing; I do not agree with that account, but because my point in this chapter is to recast already visible scholarship in terms of how individuals might use it strategically, the date works well in this case. My claim is that the computers and writing community's knowledge between these years is best represented as competing cyborg narratives, not as a single monolith, if the end goal is to facilitate and encourage acceptance of scholars' work. That is, if what has been done in the past has the potential to shape what is done in the present and what may be done in the future, then it makes sense to construct a full account of what has been done in the past. Such an account promises to enable scholars in the computers and writing community to ground their work in an awareness of previous studies, no matter how progressive and innovative their new work may be. This grounding, I believe, is an important form of acceptance for scholars, helping them see they are not alone and that their work indeed contributes to the community's growing professional knowledge about subjects of mutual interest.

LEGITIMATION AND THE NATURE OF CYBORG NARRATIVES

Because scholars in the computers and writing community have been actively seeking acceptance of their work in various contexts, they have been doing more than conducting sound research and sharing important ideas in publications and at conferences. These computers and writing scholars are writing, in many respects, for their professional lives, what Donna Haraway (1985) would term *cyborg writing*. In "Manifesto for Cyborgs," she explains, "Cyborg writing is about the power to survive, not on the basis of original innocence, but on the basis of seizing the tools to make the world that marked them as other" (p. 175). The implied writers in Haraway's description are clearly conscious of the oppressed nature of their lives and thus the political nature of their work, just as computers and writing scholars realize generally that their long-standing struggle for acceptance through writing is also highly political. In this way, the scholarship of those in the computers and writing community may be seen as cyborg writing, political activism for a changed academy in which their work and that of their colleagues will be accepted more often. This advocacy, I want to emphasize, represents an investment in individuals and the community they share, rather than any construction of computers and writing as a field or discipline. That is, we should think about scholarship as representing people—the authors and readers, of course, but also those who may follow. Additionally, we can think about the computers and writing community broadly in this way, bringing in those who contribute to only one or two of its areas and emphasizing their contributions as every bit as important as those who contribute in more areas.

The term *legitimation* offers an important way of thinking about such acceptance because it emerges from the same postmodern and postindustrial milieu as Haraway's concept of cyborg writing. In *The Postmodern Condition: A Report on Knowledge,* Jean Francois Lyotard (1984) describes the relationship between narrative knowledge and legitimation. His tack is to draw a distinction between cultural knowledge, which he represents as narratives, and scientific knowledge. In the modern era, Lyotard argues, large-scale master narratives, like the American dream or industrialization for progress, existed. Individuals and groups legitimated professional and personal choices by measuring their direction and substance against the relevant master narratives. With the rise of postmodernism and postindustrialism, however, according to Lyotard, master narratives were destabilized, resulting in a series of competing little narratives. Such change also destabilized legitimation, and it has become more context-bound, more localized to specific communities and organizations. Some scholars, like Bruno Latour (1993), in his *We Have Never Been Modern,* have problematized Lyotard's account, finding fault in his separation of cultural and scientific knowledge, but the summary of legitimation's evolution is generally upheld. For the purposes of this chapter, it works especially well because Lyotard's explanation of postmodern and postindustrial legitimation emphasizes complexity in its conceptualization of little narratives, the sort of complexity that comes from rejecting an overly simplistic, monolithic view of knowledge in the computers and writing community as this chapter seeks to do.

Upon this intellectual foundation, a definition of cyborg narrative can be constructed. First, recasting visible scholarship in the computers and writing community from 1979 to 2000 as cyborg writing enables the political capital of that body of knowledge to be apparent, thus emphasizing the way professional knowledge can support individuals strategically in their various contexts. In the body of knowledge, themes can be located, paths of intellectual development that emerge from the close and holistic consideration of scholars' perspectives. These themes, ultimately, serve as cyborg narratives, representations of knowledge against which present and future work in the field may be legitimated. They are not master narratives in Lyotard's terms, but little narratives, competing against each other for prominence, and they may intersect in any number of ways, if they do at all. What matters in the end is that knowledge in the computers and writing community be seen as more than a monolith because such a singular vision does not provide the best grounds for the legitimation of present and future scholarship. The more complex representation of community knowledge that competing cyborg narratives offer is much more promising.

Before proceeding further, let me be specific about why such narratives are so key for the computers and writing community. If we imagine the community to be a monolith, then everyone who becomes invested in the community can only assess their contributions against that monolith, an approach that would effectively de-emphasize the contributions' importance by making them seem small and even perhaps inconsequential. On the other hand, enabling those contributions to be measured against specific smaller narratives in computers and writing creates a scenario in which anyone's contributions can be understood in more specific relief, their importance being more realistically represented. In the early days of the computers and writing community, many scholars did a great deal across the computers and writing community, effectively contributing to multiple narratives, but as the community has matured, growing more diverse and more complex, it would be unrealistic to expect many to do the same. If we were to examine the career of a community member like Christine Hult, for instance, then we'd make a mistake in wondering why she doesn't do more work with creating new textual genres or developing new media technologies in industry; clearly her innovations in teaching (and especially online education) more than establish her as contributing a great deal to the community. A monolithic view would make such contributions seem less important because of their specific focus, but a little narratives view shows how vital they have been to pedagogy by enabling a little narrative on that specific area of interest and thus demonstrates why the contributions are so important, even in their specific way. We simply shouldn't ask Hult to do more; it wouldn't be realistic, and it wouldn't be fair. We should instead emphasize the good work she already does, work with more than enough breadth and depth to interest and challenge individuals for entire careers. The computers and writing community, finally, is too sizable for us to expect anyone to be well versed in all of its components, and this reality is not a loss; it's a chance for us to encourage outstanding individuals to pursue areas in it and make a difference.

In the sections to follow, I introduce two cyborg narratives drawn specifically from visible scholarship in the computers and writing community, and I demonstrate how they promise to legitimate present and future work in the community. The narratives are the following:

- Textual Transition and
- Pedagogical Evolution.

Other useful cyborg narratives are sure to exist, of course, depending on who is thinking about knowledge in the computers and writing community, so these two are intended to be examples, not held out as the only ones significant. At the same time, though, these two have been highly prominent; they are likely to appear in many accounts of the community.

TEXTUAL TRANSITION

One of the most multidimensional cyborg narratives evident in the computers and writing community's scholarship chronicles the evolution of textuality, from early projects exploring word processing and grammars of the screen to more contemporary examinations of social and cultural implications of hypertext and multimedia. Because of its range and scope, this narrative will enable the legitimation of a wide range of scholars' work in the present and future.

The narrative begins in the early and mid-1980s, when scholars first began to suggest that computers enable new authoring, revising, and editing practices through word processors. Among the earliest applications was *WANDAH,* developed at UCLA by Lisa Gerrard, Michael E. Cohen, Andrew Magpantay, and Ruth Von Blum. This application included specific attention to idea generation, as well as drafting, revising, and editing, so it implied a process model of writing for its users. The process model proves still vogue today, in fact, so present and future designers should learn about *WANDAH* in more detail to gain a sense of the challenges Gerrard, Cohen, Magpantay, and Von Blum faced and resolved. In "Studies in Word Processing," published in *Computers and Composition* in 1986, Gail Hawisher suggested that word processing was impacting writing across the disciplines, not just in composition studies. Hawisher's article showed, effectively, that work from the computers and writing community can inform many disciplines in important ways, a perspective that resonates directly for present and future scholars who will seek acceptance of their projects in and across various contexts. Despite increasingly widespread use of word processors at that time, however, doubts also existed. Isn't there something unique and important, critics wondered, about the nonelectronic writing process? Something gained in the labor of love? These questions reflect a substantive critique of word processors, and in that way, they are important. The questions are even more important, however, for what they demonstrate about the emergence of any new technology: that resistance will play a foundational role in the way the technology is seen and used. If present and future computers and writing scholars can be sensitive to and respectful of critique that will emerge, or if these scholars can offer critiques themselves that are constructive and well reasoned, then a wide range of new and carefully designed technologies important for the community should emerge.

In the late 1980s and early 1990s, scholars turned their attention to the grammars of the screen, and this attention demonstrates for present and future scholars how individuals think about and try to make sense of new technologies they encounter. Grammars, in this context, are meaning-making practices associated with particular media. In America, for instance, the printed page implies a top-to-bottom and left-to-right reading convention, whereas traditional Korean requires vertical reading from right to left. Scholarly debate about grammars of the screen focused on several central questions: Does reading a text on a computer screen imply different conventions? What specifically are these different conventions, if they exist? Likewise, what does their possible existence mean for authors who would have their texts read on screen? This discussion points specifically to the computer screen as the technology under view, but any alternate display technologies would see variations of the same questions. They are indicative, I believe, of the way individuals come to know any technology that has the potential to transform core literacy values. Two chapters from *Critical Perspectives on Computers and Composition Instruction* (1989) offer an illustration of scholars' attempting to respond to such questions in the late 1980s and early 1990s. In "Redefining Literacy: The Multilayered Grammars of Computers," Cynthia L. Selfe argues that new grammars are indeed inherent in screens; she theorizes, more, that these new grammars reflect "layers" of literacy, bringing a key connection between the printed page and the computer screen. The term Selfe employs for this phenomenon is "*computer-mediated literacy*" (p. 3). A different look at grammars comes from Christina Haas (1989), who, in "Seeing It on the Screen Isn't Really Seeing It: Computer Writers' Reading Problems," argues that reading texts on the screen requires fundamentally different composing processes. Haas suggests that the screen does not enable readers and writers to gain a sense of the whole text like holding a book would, and her perspective clearly establishes that different grammars do not always mean better grammars, a reality as true today as it was then.

The narrative of textual transition continues in the early and mid-1990s with the emergence of electronic hypertext, sparking a debate about whether it could be seen as a distinct and new genre of text. In the future, scholars will no doubt introduce many new textual forms as potentially unique, so understanding the debate around hypertext promises to enable these scholars to anticipate the field's response. More, the scholarly activity around hypertext shows what a complex and detailed process such a claim necessarily stimulates. It is important first to identify *electronic hypertext* as the subject in this paragraph, not *hypertext* alone. The term *hypertext* was coined in 1945 in "As We May Think," an *Atlantic Monthly* article by Vannevar Bush, and it has been shown to have as much applicability to novels and short stories as to electronic media. Electronic hypertext, then, refers specifically to nodal and multilinear constructions of text enabled by electronic media, like *Storyspace,* CD-ROMs, and the World Wide Web. Perhaps the earliest widely known representation of electronic hypertext in the computers and writing community was Jay David Bolter's term, *electronic writing space*

(1992), by which he was describing hypertext's potential to offer new writing and reading opportunities through linking and multiple nodes. Bolter's term was followed in the mid-1990s by Michael Joyce's *constructive hypertext* (1995), which emphasized the way readers of electronic hypertexts may likewise be seen as authors because they select which hypertext links they follow, thereby encountering different texts. Johndan Johnson-Eilola, with his *Nostalgic Angels: Rearticulating Hypertext Writing* (1997), offers the broadest and most satisfying discussion of hypertext in the computers and writing community's scholarship thus far, connecting it with ideology and economy. No matter where the future of computers and writing goes, it will necessarily remain connected to critical issues like those Johnson-Eilola raises. It makes sense, then, that considering his view will be important for future and present computers and writing scholars alike, as will looking at scholarship by Joyce, Bolter, and others. The whole history can inform the way genres of text emerge and are shaped.

During the early and mid-1990s, scholars began thinking about unique features of other electronic media, like e-mail and MUDs. The introduction and adoption of these media show how new media may be received, enabling scholars to anticipate both challenges and support. One of the earliest books in the computers and writing community to address e-mail and network technologies was Carolyn Handa's *Computers and Community: Teaching Composition in the Twenty-First Century* (1990). It shares a number of social perspectives on pedagogy, demonstrating that networked media enable associated texts to serve as sites of interaction in teaching and learning. This social view continues to be prominent today, in fact, so present and future scholars should know both to look as far back as Handa's book and to look at what followed it in field discussions. In a 1993 article, "Electronic Mail and the Writing Instructor," Gail E. Hawisher and Charles Moran call for a large-scale rhetoric of e-mail, moving beyond simply examining specific e-mail conversations for discursive innovation and for unique modes of dialogism. Their model calls for close study of ". . . genres, audiences, voices, uses, and the extent to which any and all of these are influenced by the properties of the medium" (p. 629). To this date, Hawisher and Moran's call has not been fully realized, but their challenge implies questions that should be central for the introduction of any new medium. That is, after scholars have explored the grammars associated with a new medium, if appropriate, their attention will likely turn to rhetoric and critical analysis, helping push understandings of the medium forward in careful and responsible ways. E-mail has also emerged prominently in popular culture, and the connection between that spectrum and the computers and writing community seems rich with promise. A recent film was even titled *You've Got Mail,* for instance, the words of the title being the America Online sign-on voiceover for users who have e-mail waiting for them. With such attention in popular cultures comes, I believe, a degree of acceptance for the computers and writing community's work, whether manifested in interview opportunities with newspapers and magazines, colleagues coming down the hall for a chat about a particular medium, or just simply the knowledge that colleagues and friends will understand references to the medium in conversation. Present and future computers and writing scholars, it follows, should think carefully about how new media proliferate both in and beyond the field, and e-mail offers an important example because its own developmental path has been so broad and varied.

MUDs also emerged prominently in the computers and writing community during the early to mid-1990s, offering present and future scholars a chance to see how a second medium became accepted as well. Thinking about the emergence of e-mail and MUD technologies together, in fact, should enable scholars in the computers and writing community to gain a more holistic sense of the way media are received and integrated in the field. MUD, as readers may know, is an acronym for Multi-User Domain or Multi-User Dimension, and as a technology, it operates essentially as a virtual reality world that facilitates real-time interaction and enables players, as they are called, to construct worlds of their own. In the computers and writing community, a specialized type of MUD, called MOO, has been most popular, and it emphasizes objects players can make, alongside the interaction and world construction of MUDs. In MOO environments, players encounter multiple and rapidly scrolling threads of text, and what generally occurs first is that new players feel overwhelmed, a sort of resistance to MOO textuality. Such resistance does not mean that players are not trying hard or that they do not know what to expect; it simply means that adjusting to the conventions of MOO is challenging for some. Present and future computers and writing scholars, in fact, should take note of the way resistance emerges as part of any new medium's rise to prominence in the field. Many times, those who feel unsure about a medium can articulate concerns and worries that are useful for the field and for designers specifically. As players MOO more and more often, the multiple threads of text do read more easily. It's not, finally, that the players read more quickly; it's that they develop what might be termed an *individual meta-reading heuristic,* where they scan screens for interesting words or phrases or perhaps the comments of other players with whom they are friends. This development also provides important information for present and future scholars because it stipulates a process by which individuals new to a medium can come to know it. MOO as a medium in particular suggests that resistance emerges first, to be followed by small strides and then a more sophisticated degree of use, and these stages, I believe, should be characteristic of the way future media come to be known and widely used as well.

The next stage of textual transition emphasizes *multimedia,* the term here referring to the electronic presentation of several digital media, like graphics and audio and video clips, as well as more traditional print formations, like paragraphs and sentences. Because multimedia is such a broad term, the way it has come to be integrated into the computers and writing community proves especially instructive for present and future scholars, as their own innovations will likely rely on more advanced iterations of at least some of the same media. As with individual media introduced earlier in this chapter, the rise to prominence of multimedia involves both resistance to it and careful progress with it. Advocates of multimedia technology today suggest that its integration into education means a new and unique generation of "wired kids," able to think about and interact among many media in a short period of time, but unable to sustain a more singular focus in educational contexts. These wired kids, so the thinking goes, are a vision of the future. On the other hand, critics of multimedia worry about these wired kids, believing their education is leading them away from the discipline and determination that the future will require. Those who resist multimedia often suggest it is for entertainment, instead of education. In "HyperRhetoric: Multimedia, Literacy, and the Future of Composition," Gary Heba (1997) articulates a middle-ground position, willing to allow that multimedia may offer an important new educational technology, but unwilling to consider it without critical interrogation of its nature and impact. This hopeful but careful approach should prove useful for scholars thinking about any new technology or media, as it emphasizes thought and assessment alongside enthusiasm. One of the most important scholarly examinations of multimedia for the computers and writing community today is Janet Horowitz Murray's *Hamlet on the Holodeck: The Future of Narrative in Cyberspace* (1997). Citing an impressive range of projects and relying on her own experiences at the Massachusetts Institute of Technology and other institutions, she reads multimedia into the larger narrative genre, etching a place for it in much broader discussions of textuality. Such work, when it can be done, forges important connections between generations of scholars interested in texts and technology. Future scholars might, for instance, think about the way media in their lives enables correspondence and then look back at today's use of e-mail for keeping in touch, followed by studying epistolary exchanges in literature, whether novels like Samuel Richardson's *The History of Pamela; or, Virtue Rewarded* (1794) and Bram Stoker's *Dracula* (1932), or classical era exchanges between Socrates and students. Such an informed sense of any medium should prove valuable for future scholars, as it does for us today.

The final stage of the cyborg narrative of textual transition emphasizes the way globalization has resulted in more diverse textual production and consumption practices. These practices promise to continue to be prominent in the computers and writing community, so scholars should know the work that has already been done in order to make well-grounded choices in the future. The Georgia Institute of Technology's Graphics, Visualization, and Usability Center (GVU) has conducted a number of World Wide Web user surveys in recent years, and their 1999 numbers were indicative of an increase in diversity: more and more women online, a growing number of users outside of North America, and an increase in users from poor and disadvantaged backgrounds. These numbers are not enough, however. That is, what present and future scholars should take from the Georgia Tech research is not that diversity is increasing, although that is important, but that it is essential to be attentive to diversity at all times. Gail E. Hawisher and Cynthia L. Selfe's *Global Literacies and the World Wide Web* (2000) presents a series of collaborations undertaken by scholars in the United States and around the world, and although the book demonstrates impressive geographic diversity, its larger and more important project is to suggest that new literacies are being constructed on the World Wide Web. These literacies would not otherwise exist, which underscores the importance of the computers and writing community's work with global collaboration, and their existence further indicates to present and future scholars that they must look carefully for new and important textual forms or literacies in the virtual spaces that they create and inhabit. Although we cannot from the present imagine what the future will involve, we can say that a future attentive to diversity promises to be most conducive to careful and responsible textual transition.

Into the future, both print and electronic texts will continue to be prominent, as the narrative just sketched demonstrates. Scholars need to understand that their choice is not between all computer technology and no such technology. Instead the challenge for the present and future both must be to examine whatever textual production and consumption practices prove most important, from those rich with computer technology to those without any. This vision ensures that a wide range of scholars' work can contribute to the computers and writing community's growing knowledge about print and electronic textualities.

PEDAGOGICAL EVOLUTION

The second narrative outlined in this chapter is the cyborg narrative of pedagogical evolution. In the following discussion, teaching and learning are conceived broadly, representing a full range of pedagogical innovations and experiences. Most important about what follows is the amazing range. In only the 20 years between 1979 and 2000, the computers and writing community has seen variations of independent, software-specific, networked, and distance education, as well as more traditional classroom teaching and learning. This diverse path of development should encourage present and future scholars in the computers and writing community to be innovative and to see that the choices they make connect directly with those made by earlier scholars.

In the 1970s, when the cyborg narrative of pedagogical evolution begins, much of the writing software available was written specifically by writing instructors, this simply because no other options existed. Although it is unclear if educators in the computers and writing community will do such work again in the future, this experience clearly will prove vital for guidance and encouragement, if such work is indeed taken on. One of the most interesting aspects of 1970s and early 1980s software is the way it reflects a host of different pedagogical values. Some early applications, like those in *The Writewell Series,* positioned students as independent learners. A student would, following this model, sit down at a computer terminal and be led through a series of activities on punctuation, voice, or whatever the learning subject is for the day. Other applications, such as *Writer's Workbench* and *Epistle,* provided editing assistance for students, helping them to review their papers carefully before submitting them to be graded. Perhaps the best known early software innovations in computers and writing are those of Hugh Burns: *BURKE, TAGI,* and *TOPOI,* each designed to assist students to think through paper topics by asking questions of rhetorical significance. Yet another pedagogical variation demonstrated by software may be seen in early wordprocessors, like *WANDAH,* which focused on enabling students to input their ideas into a computer file that could be saved and retrieved later when a writer wanted to make revisions or edits, as well as to generate ideas to be expanded and discussed in the paper. Present and future scholars should also appreciate that 1970s and early 1980s applications were not smoothly designed in today's terms. Early word processors, for instance, were not like contemporary word processors with carefully designed user interfaces and command structures that make operation user friendly; instead, these word processors often required specialized commands, and the interfaces were complicated.

During the early and mid-1980s, pedagogy and software continued to be closely associated, and corporations entered the mix. With so much of today's software for teaching and learning authored by corporations, and with the future promising all the more, it makes sense for present and future scholars to have a sense of these earlier times as a way of grounding their decision making and assessment of software options. With larger budgets and more extensive resources, corporations made significant changes to software in the early 1980s, especially in making software more user friendly. Applications like *Applewriter* and *Wordstar,* which operated much more simply than earlier word processors, well demonstrate this transitioning. More, they show just how much difference corporate funding and support can make in only a short time. Today's corporations range from giants like Microsoft, Apple, Compaq, and Hewlett-Packard, to smaller local organizations, and each necessarily shapes the way scholars in the computers and writing community work, whenever its products are purchased and integrated. The future will likely be equally reliant on corporations to provide sound software options, for better or worse, so scholars should think carefully before making choices. In the early and mid-1980s, the move toward corporations for hardware and software also began to raise critical questions about pedagogical values. Should academic institutions, critics wondered aloud, teach students to be "good little consumers" of technology products? In what ways might consumerism and corporations' growing need for technically skilled employees influence curricula? Richard Ohmann's "Literacy, Technology, and Monopoly Capitalism" (1985) is perhaps the most often-cited early publication in the computers and writing community to raise such critical questions. In particular, he brought attention to economic and class issues associated with the shape and substance of curricula. The sorts of issues Ohmann raised continue to be important today, as they should into the future. Currently, for instance, the open-source software movement has gained prominence as a response against the licensing and trade monopolies of corporations, and teachers are left with particularly difficult decisions because choosing open-source software may be the right option politically for them, but may also require them to devote significant time to learning new applications. The open-source movement should be important, I believe, for future scholars too, as it continues to develop, and understanding the early interweaving of pedagogy and software in computers and writing should help scholars make more informed decisions.

By the late 1980s, the computers and writing community's pedagogical evolution began to grow in diversity, featuring scholars in literary studies and other fields. Scholars doing work in the present and future should be especially mindful of these early experiences because they show how interdisciplinary connections can be forged through pedagogy and correspondingly how the community can grow in important ways through diverse contributions. Few sites of computer-rich pedagogy have been more influential than the literature classroom, and one of the earliest scholars to introduce computers into literary studies was George Landow. He and his Brown University students developed a hypertextual guide to Victorian literature called the Victorian Web. At that time, the World Wide Web had not yet been introduced, so Landow's vision is all the more impressive. His classes used UNIX-based *Intermedia* software for their design, and its function, more or less, was to enable files on a server to be conceptually and functionally linked to each other such that multiple users on multiple computers could view and navigate them.[1] The *Intermedia*-based Victorian Web is demonstrated in the following screen capture:

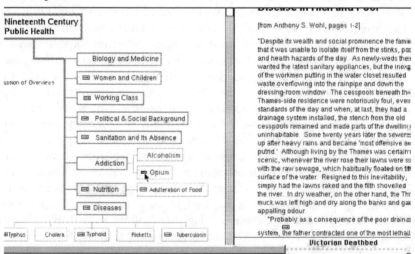

---

[1] It should be noted here that the Victorian Web is still alive and being updated regularly by Landow and other scholars around the world. As of this writing, readers may find the Web site at http://landow.stg.brown.edu/victorian/victov.html. Also of note, the Victorian Web has seen several technological changes: from *Intermedia,* as described, to *Storyspace,* and then to HTML for display on the World Wide Web.

In the left-hand window of the screen capture, a schematic of the Victorian Web's files, or nodes, is located; users selected topics of interest, and with each choice, they were able to see and select more specific topics, eventually locating desired information. The right-hand window displayed the specific information requested, in this case a discussion of Victorian disease. In looking at the Victorian Web, readers can no doubt see the way that early work with *Intermedia* connects to today's pedagogies grounded in web publishing. Present and future scholars should know about such early pedagogical innovation to help them appreciate the way their unique teaching and learning experiences link to those of the past.

One of the most profound moments in the computers and writing community's pedagogical evolution occurred in the late 1980s and early 1990s, when network technologies were brought into classrooms. Since that time, networks have continued to grow in prominence as technologies for teaching and learning, and they will no doubt be critical for the future of pedagogy in the computers and writing community as well, making an understanding of earlier experiences especially important. The computer networking first introduced in the late 1980s was termed a *local area network,* or LAN, and it essentially involved several technological components: a modem in the source computer that allowed it to convert information into a transferrable format, a wire between all computers to be involved in any such transfers, a central data transfer router, and a modem in the target computers that allowed them to transfer data they receive back into a readable and usable format. A LAN does not enable outside computers users to participate in information transfers; only specific computers are selected, and individuals have to be at these computers to access the LAN. One of the earliest projects using network technologies to support innovative teaching and learning was Electronic Networks for Interaction, or ENFI, first implemented at Gallaudet University but rapidly expanding to other institutions at that time. Its designer, Trent Batson, worked to develop ENFI to encourage collaboration and a redistribution of classroom authority. In 1990, following use of LANs in their writing classrooms, Thomas Barker and Fred Kemp (1990) published "Network Theory: A Postmodern Pedagogy for the Writing Classroom" as a chapter in *Computers and Community: Teaching Composition in the Twenty-First Century.* They also report promising advances in collaborative pedagogy as a result of their use of network technologies. Kemp, in fact, was one of the authors of the original *Daedalus Integrated Writing Environment (DIWE)*, LAN-based software that enabled users to share documents and interact with each other about them. In its latest release, *DIWE* is web-based, no longer relying on LANs, so it demonstrates a particular sort of pedagogical evolution around network technologies. Present and future scholars who come to know about *DIWE*'s growth can think about the way their own teaching practices may lead to a reshaping of software. And the early contributions of innovators like Batson, Kemp, and Barker promises to continue to inform pedagogy into the future, as more and more educators in the computers and writing community think about the way networks enable new possibilities.

Building on important lessons learned with LANs, scholars in the computers and writing community next began to employ multiple electronic media in pedagogy, especially including electronic mail, the World Wide Web, and MOOs. This diverse stage of pedagogical evolution continues today, demonstrating that it is particularly valuable for present and future educators to know as a way of grounding their work. Hawisher and Moran's "E-mail and the Writing Instructor," cited earlier in this chapter, is one of the most mature early studies of e-mail. In this article, the authors develop a broad approach to e-mail, thinking about ways a rhetoric of the medium might be developed. Cynthia L. Selfe, in her "Theorizing E-mail for the Practice, Instruction, and Study of Literacy" (1996), even suggests that e-mail can be used to develop literacy practices, forging a key connection between print and electronic media. Educators in the computers and writing community, then, seeking to encourage writing and communication excellence in their classes, can utilize electronic mail in conjunction with print media, with each being highly valuable. E-mail resembles any number of asynchronous electronic media, and it's important for scholars to appreciate their variety as well. Examples of these media include bulletin boards and newsgroups. Boyd H. Davis and Jeutonne Brewer's (1997) *Electronic Discourse: Linguistic Individuals in Virtual Space,* in fact, presents a discourse analysis of a bulletin board created by the authors on which students interacted about primary newspaper texts assigned as class reading. It represents a mature research approach, one that well illustrates how asynchronous electronic media have been intertwined with contemporary teaching and learning. Newsgroup technology also provides an apt forum for many wide-ranging discussions. Just like an e-mail discussion list, newsgroups are defined by posts that computer users submit asynchronously; the only difference is that newsgroups can only be read in threads, whereas discussion list participants can see posts in their e-mail accounts or in archives. The most important outcome of previous work with e-mail is perhaps not any of its specific pedagogical applications, but the way it pairs with other asynchronous media to offer educators multiple opportunities to challenge students.

Like e-mail, the World Wide Web opens up a range of pedagogical options, and today's integration of it into the computers and writing community provides a valuable foundation for present and future educators who will use it. First and foremost, the Web brings opportunities for electronic publishing, both for teachers and students, and this feature has a significant impact on classroom pedagogy. Teachers can, for instance, engage students in rich and substantive dialogues about audience and then ask students to apply their new understanding in designing and authoring a document for the Web. Interestingly, Web authoring also returns the computers and writing community to debates about software development—if teachers should write software, how much technical knowledge teachers should have. The crux of the debate centers on hypertext mark-up language, or HTML, which is not a heavy or difficult language, but nonetheless requires knowledge of file systems and programming tags. Software options, like

Netscape's *Composer,* Macromedia's *Dreamweaver,* and Microsoft's *FrontPage,* have become available, and these enable users to author web pages without knowing HTML. In general, advocates of HTML-level knowledge continue to find fault with the extra or operationally irrelevant tags web-authoring software like *Composer* and *FrontPage* adds, as well as arrangement of the HTML code, and advocates of the software believe that teachers and students should use the most user-friendly option, that the software is an evolutionary stage beyond HTML authoring. The debate will soon be moot, however; applications like Dreamweaver write clear code, not the problematic code created by Composer, FrontPage, and other like applications, so both sides of the debate will be able to win: clean code and advanced software all at once. One project that addresses and expands such existing practice-based tensions is Sibylle Gruber's *Weaving a Virtual Web: Practical Approaches to New Information Technologies* (2000), in which authors locate critical discussions of the World Wide Web in a series of pedagogical approaches. Gruber's book is but a beginning, however, as scholars need to look more and more at the pragmatic applications of technologies like the Web. Such work proves a viable and interesting challenge for present and future educators to assume.

MOOs also have a profound impact on pedagogical evolution, impact that promises to continue into the future. Indeed present and future scholars in the computers and writing community working with any variation of synchronous media can benefit from looking at the community's MOO scholarship and activity. One of the earliest educational MOOs was MediaMOO, developed by Amy Bruckman and colleagues at MIT's Media Lab.[2] MediaMOO was the first home of the Netoric Project, a virtual organization that sponsors weekly MOO meetings for interested members of the computers and writing community, and MediaMOO was partially profiled in Stephen Doheny-Farina's (1996) *The Wired Neighborhood.* Since then, other educational MOOs have emerged, including Tari Fanderclai's Connections, which is the current home of Netoric; Cynthia Haynes and Jan Rune Holmevik's LinguaMOO, which serves as the virtual home of important new projects like QueerMOOnity; and Diversity University, which holds weekly meetings for educators across the disciplines interested in incorporating MOOs into their teaching. Seemingly a straightforward technology, then, MOOs reflect significant diversity, diversity that directly informs pedagogy. Scholars requiring students to MOO might, for instance, select LinguaMOO, if they want to use that MOO's web interface, instead of asking students to download specialized

---

[2] In this chapter, a distinction is made between "educational" and "social" MOOs, and this separation is common in scholarship about MOOs. An *educational MOO,* more or less, is one where its chief mission is to support teaching and learning opportunities. A *social MOO,* on the other hand, operates differently, with its first priority being to facilitate social opportunities for players. These characterizations overlap at times, of course, but they prove useful to examining the fundamental ethos of these virtual environments, especially as contextual knowledge becomes important to understanding actions and opportunities in them.

software, or if they want to be able to frame in hypermedia to the MOO experience. On the other hand, scholars wanting to emphasize text-based description and activity would likely select Connections, where a web interface is not available. The distinction is technical—enCore as an operating system for Lingua versus Jay's House Core (JHCore) for Connections—but it is also uniquely pedagogical, and present and future scholars should note carefully the way educators have navigated the options available to them. Thus far, the best known scholarly project to take on such issues is Haynes and Holmevik's (1998a) *High Wired: On the Design, Use, and Theory of Educational MOOs,* which actually includes an enCore database system with which readers can install and operate their own MOOs. *High Wired* authors come from a range of disciplinary backgrounds, so demonstrate the way work in the computers and writing community can extend well across the curriculum, and they also describe a variety of pedagogical innovations that can be useful for present and future scholars, from the way theatrical performance can be conducted in MOOs, to the way writing centers may use them to support student writers.

MOOs have also brought new excitement to specialized communities within the computers and writing community, and this aspect of pedagogical innovation proves particularly important for present and future educators who should know their work does not have to impact all of the computers and writing community at once to be highly valuable. Key writing center innovators with MOOs, for instance, include Eric Crump, whose *High Wired* chapter (1998) recounts his MOO tutoring and tutor training experiences at the University of Missouri's Online Writery, and Jennifer Jordan-Henley and Barry M. Maid (1995a), who have written about their tutoring collaborations for *Writing Lab Newsletter* and *Computers and Composition* (1995b, 1995c), as well the book collection *Taking Flight with OWLs: Examining Electronic Writing Center Work* (2000). Maid, then at the University of Arkansas at Little Rock, trained his graduate tutors in rhetoric and composition to interact in MOOs, and they tutored Jordan-Henley's composition and technical writing students at Roane State Community College. Maid and Jordan-Henley together provided leadership and guidance for both groups. MOOs have helped other specialized communities to enter into conversations about pedagogy in the computers and writing community as well. Bruckman's (1997) dissertation project, for instance, titled *MOOse Crossing: Construction, Community, and Learning in a Networked Virtual World for Kids,* developed a MOO for elementary school children, helping expose them early to computer technologies, and she continues its work today in her faculty role at Georgia Tech. Thinking about the pedagogical diversity associated with MOOs, present and future scholars should see that their own teaching activities can be broad and that even very specialized or local variations prove important to the field.

The last stage of the narrative of pedagogical evolution emphasizes distance education, which has grown in amazing fashion from the mid-1990s to today. This growth will no doubt continue into the future as well, making an understanding of its history highly important for present and future educators alike who will need to make careful decisions about the way they interact with it. It's feasible even that many faculty positions may evolve into distance models. The idea of distance education is much older that the 1990s, it should be noted first, whether one begins with the videotape-by-mail model of the mid- to late 20th century, the early-20th-century distance-learning innovation of the University of South Africa, or even with epistolary exchanges between students and teachers in the classical era, but the specific influence of computer technologies is a recent phenomenon. In its current form, distance education offers access for students who could not study in residence at a college or university. Utah State University and Texas Tech University, for instance, offer online master's degrees in technical communication, and their programs serve as models for those who would seek to reach adult graduate populations. Featuring students on all seven continents, the British Open University is perhaps the world's distance education leader, using a range of digital media to enable learning, so it also serves as an excellent example. At the same time, states have developed comprehensive distance education programs, and these prove innovative and appropriate as models as well. The University of Alaska system, for example, makes extensive use of video conferencing and computer technologies to reach remote students. Finally, distance education has brought a return to the sorts of critical questions Ohmann first posed in the 1980s, as education in general is becoming more and more intertwined with corporations. The Michigan Virtual Automotive College, one example of note, led by former University of Michigan President James Duderstadt, reflects a collaborative relationship among Michigan, Michigan State University, the State of Michigan, the United Auto Workers (UAW), and the automotive industry, among others. More, organizations are designing unique educational programs; their future at this point remains to be seen, but their presence is clearly being felt. Having a sense of the way distance education has emerged in the computers and writing community should enable present and future scholars to make sound, informed choices.

Ultimately successful teaching and learning will involve both traditional and progressive options—with and without computer technologies. The years from 1979 to 2000 have seen a considerable range of pedagogical evolution, and the future promises to be equally dynamic. Future scholars who come to know the contributions already made by individuals in the computers and writing community can see that their work contributes meaningfully to a rich and diverse base of knowledge.

CONCLUSION

This chapter's discussion of cyborg narratives began with the idea that computers and writing scholars often struggle for acceptance of their work in the cyborg era, and I suggested one useful way to proceed would be to recast the community's already visible scholarship not as a single story, but as a convergence of such competing narratives. Against these narratives, present and future work in the computers and writing community may be legitimated; they are a technology, in essence, for acceptance. Perhaps even more importantly, they help us emphasize the important contributions of many in the community, whether they do work in multiple areas or a single one. At the same time, the cyborg narratives of textual transition and pedagogical evolution will not prove themselves to be a panacea. It is important that computers and writing scholars continue to think about the substance and implications of these two narratives, and it is equally if not more important that new narratives are proposed, alternate threads scholars see in the work of individuals in the computers and writing community. We must keep thinking and writing, keep emphasizing how our work can make a difference for everyone in the community.

Community Voices

Donna Marie Jarma

How did you come to be active in the computers and writing community?

Throughout my thirty-one years of teaching, I have watched the computer become a more integral part of the classroom scene. From the seniors I teach at the high school level to my dual credit college classes to my prospective teachers class to my own doctoral work, the computer has allowed me to create and be part of communities of communication and classwork. My high school students not only research on the computer, but also complete assignments centered around computer journaling. My Issues in Education classes handle Early Field Experience through shared observations and feedback in the small community of prospective teachers. Finally, my own doctoral work has taken me into the MOO and onto learning spaces such as Blackboard and WebCT, not to mention a variety of listserves and general e-mail exchanges. In short, the computer creates communication opportunities on a regular basis. As a career English teacher, I welcome the community atmosphere created by each unique gathering of voices. In each instance, I have seen the group develop outside assignments into the realm of personal communication. I see so much potential for so many aspects of the writing process as well as simple, human interaction.

Cynthia Jenéy

What worries you about the computers and writing community, and why does it worry you?

What alternately worries and amuses me is the astounding lack of innovative and radical thinking I see in both online and fleshmeet circles focused on computers & writing. This is a time when innovation and upheaval should be our goal. Replications and repetitions of old methods inside electronic boxes, and "new" techniques that have been around for 20 years threaten to drag us into even more stifling practices than ever before. Computer technologies are not designed to erupt with organic bursts of brilliant expression: they are specifically engineered for control, structure, collection, and organization of one of the least useful concepts of writing: "information." Too many of us have been sucked into the Elbovian/Rosean morass of the nurturing composition teacher; we're handholding and guiding our students like lambs into the humming box, rather than assuming our ancient duty as fearless facilitators of symbolic action mayhem. The transcontinental neural net is out there, its nodes practically bleeding with the propinquity of global humanity and we show them "http://stuffyoucanuseforatermpaper.com".

The best ideas are on the fringes, the best people are lobbing theoretical neutron bombs at the status quo, the best studies are way outside the disciplines of composition and pedagogy, and the best student writers in our classes see a lot of our "writing with technology" lessons for what they are—a sellout (or a buy-in?) to the companies who market the gadgets and the containers to our admins and IT directors. While 80% of us are fussing and crying over downloaded, plagiarized, and just plain badly written work, a small number actually have figured out that the technology isn't the point. Good writing is still about ripping a powerful phrase into the guts of the illuminati and coming out reeking of their awed disillusion (I would stand by this claim for the technical writer as much as for the poet). And even though the Internet has intriguing possibilities for writers, most of the time we're too busy fumbling around in the smoke and mirrors of graphical user interfaces and software "features" to rope the intellectual maverick. We need to bulldog the classroom machine to the ground at least once a week, or if it won't crash, or if it's too scary for some, then we at least should perform the Manganas a Caballo, bringing the machine gracefully, submissively to its knees. There should be no mercy for the machine. Not ever. I am worried that we care too much about the machine and our own ideas of the machine, about our own acute case of Future Shock, and not enough about how we can push everything to the edge of our narrow universe and see if it tumbles off into a beautiful, shattered pile of ontological junk we just don't need, never needed. It also amuses me, the way we drag so much old baggage into the experiment. We're clinging too tightly to the fence. It's time to grab that bull and rassle.

Sara D. Jenkins

What scholarly project in computers and writing has been most influential for you, and why has it been so influential?

Unfortunately, I must hedge and say that two projects of the computers and writing community have been extremely influential to me. Fortunately, the two projects are intertwined: the Computers and Writing Online conference and the Graduate Research Network Online. The online conference allowed me to get involved with research and scholarship when I could not attend the flesh-and-blood conference; such an option demonstrates that the computers and writing community is working for the best interests of the scholars who comprise it. Also, the Graduate Research Network Online allowed me to workshop a research idea with people who are actually working in the field; being able to get feedback from scholars who are currently active in computers and writing is an extremely valuable resource. Other students at my university were surprised and not a little envious when they found out about this particular computers and writing offering, and I felt even more grateful for the real sense of community that the people of computers and writing work to foster.

Billie Jones

What worries you about the computers and writing community, and why does it worry you?

Many of us are accidental technology enthusiasts—initially drawn by the lure of technologies' bells-and-whistles (or its .wav files and blogs), but we remain here because of the active learning made possible in computer-enhanced environments. As our needs (or those of our students) have led us, and as our hardware and software have supported us, our abilities and practices have developed along vastly different planes. I worry that the haves and have-nots among the computers and writing community—those who have physical and intellectual access to cutting edge technologies and those who are working with more "trailing edge technologies"—will separate into two distinct communities, resulting in a loss to both groups.

If a technologically-abled elite forms within the computers and writing community (more likely from natural division, rather than from a purposeful one), we may stop learning from one another, which may discourage those working on/with "trailing edge technologies" from seeking the growth that had once joined us as a community. Furthermore, as the cutting edge elite becomes further distanced from the "trailing edge," they may lose touch with the mainstream, from which most community members and our students come. I am happy working with "trailing edge technologies" (occasionally even at the cutting edge of "trailing edge technologies"), but I don't want to be separated from those doing cutting edge work. I believe I have much to learn from them, and I believe that by keeping in touch with the "trailing edge," their work stays rooted in the needs of our students.

Amy C. Kimme Hea

What's the best lesson you've learned from the computers and writing community, and why is it the best?

The best lesson I have learned from being a member of the computers and writing community is to continually reflect upon my teaching practices. I attempt to live up to the tradition of critical reflective practices emphasized in the work of my fellow computer compositionists. At the risk of omitting any one scholar that I admire, I want to speak generally about the dedication of members of the computers and writing community to teaching and placing their research emphasis clearly on the goals of better informed, more politically aware, and culturally situated learning in the computer classroom. As both the range of technologies in our classrooms and even what constitutes the "classroom" changes, scholars in computers and writing seek out new ways of developing pedagogy, supporting student learners, and challenging one another to grow as pedagogues. These kinds of commitments to teaching and learning encourage me to pursue rigorous scholarship, speak passionately and critically about my teaching, and remain open to new lines of inquiry and changes in our field. As I continue my work in the community of other computer and compositionists, I will remain committed to this important lesson in critical exploration.

Judith Kirkpatrick

Why do you choose to be active in the computers and writing community?

Aloha. Finding like-minded, wired colleagues from disparate venues affords me the confidence to act in my local community of faculty in the University of Hawai'i system. Without the confidence gained in mailing list exchanges and MOO professional discussion sessions, I would not have had the foundation for the positions and the leadership I have taken in promoting the advantages of online learning and teaching writing through computers. The past ten years has shown exponential growth in the potential of computers and writing, and I have chosen this pathway as where I want to be, not what I "have" to do. I enjoy change and challenges. Keeping up with the new scholars in the field gives me more than a few opportunities to reexamine how and why I ask students to do anything. Currently, I am working on a compelling project, taking what I have learned to a new community, a technology center in the middle of a subsidized housing project via a WorldCom grant. Stay tuned to this new challenge of making improvements to combat the expansion of the Digital Divide between the "haves and the have-nots." A hui hou.

Karla Saari Kitalong

Why do you choose to be active in the computers and writing community?

CHOOSE

I didn't choose the computers and writing community—it chose me. I was studying for my master's degree with Cindy Selfe at Michigan Tech when I presented my first paper at C&W in Austin, around 1990. Fred Kemp announced the availability of the fledgling Megabyte University e-mail list (MBU-L) and I joined as soon as I got home. If I hadn't felt a part of that community before, I did as soon as I got on the list. In many cases, the most authoritative and articulate voices turned out to be graduate students, adjunct faculty, or technology support staff— the traditionally marginalized members of the profession—although tenured faculty were in the mix, too. They were debating issues of importance to me, such as how to recruit, train, and retain excellent technical support staff; how to define, teach, and theorize about critical technological literacy; and how to balance tools with concepts in teaching this or that course. Plus, it was fun. I've always believed that work should be fun.

COMMUNITY

I can't remember what I said in my first post to MBU-L (I wish I could), but I remember the early fumbling as the community wrote itself. It wasn't always pretty; for example, several prominent members accidentally posted embarrassing private messages to the forum, and a key player left in a huff after a heated argument. But there was also a lot of healthy brainstorming and productive wrangling over issues. The community's evolution—inscribed in its lists, conferences, and MOO sessions—has been fascinating to watch.

ACTIVE

I took the C&W community for granted when I was at Michigan Tech, where I worked with Cindy and Dickie Selfe and a host of graduate students whose interests both reflected and helped to shape that community. I had been active in the community—in a semi-lurker sort of way—since those early MBU days, yet until I finished my doctorate and moved to Florida I hadn't really understood the community's importance to my work. In this new place, I found, for example, that it's now necessary for me to explain and justify pedagogies and ways of working that I thought were self-evident. Listserv participation, for some of my colleagues, is a nuisance and an intrusion. No one I work with (to my knowledge) has ever been to a MOO session, although some express interest in using a MOO for teaching purposes. Most of the technology work is taking place not in the English department, but in places like education, health and public affairs, and engineering! I've been forced to be more articulate about what I value and believe. This kind of critical self-reflection may make it into a public forum, but it's also active if it merely weaves itself quietly into my work.

Michael Knievel

What's the best lesson you've learned from the computers and writing community, and why is it the best?

The best lesson I have learned from the computers and writing community—and I think one learns and re-learns this every time one opens *Computers and Composition* or attends the annual Computers and Writing Conference—is to be open, accepting, and tolerant of people and ideas. I look back at the history of computers and writing, and it is characterized, above all, by open mindedness and its attendant excitement. Those responsible for establishing the trajectory of the discipline charted a course that emphasized inclusion when they so easily could have chosen to be exclusive.

What we are left with today is a community that embraces most anyone interested in computers and literacy. Unlike many academic communities, we are interested in the both the community and in industry. We readily embrace the view of our field as interconnected not only with other academic disciplines but with culture, politics, and the world. And, most importantly, the central mission remains: to help people use and understand the implications of computer use in communication and writing instruction, however that might be defined. Anyone interested? You're welcome to be a part of it.

Josephine A. Koster

How did you come to be active in the computers and writing community?

I got my first e-mail account in 1989, and slowly I learned to use tools like gopher and ftp. I began doing e-mail and fax consulting with a lot of technical clients like Bell Labs—professionals who used computers comfortably but needed my help to write. But I didn't really see the possibilities of connecting computers to writing instruction until 1993. That autumn I broke my arm quite badly; my thumb and fingers were paralyzed. Fighting to regain the use of my writing hand, I began to use a trackball for physical therapy. It seemed logical to hook it up to a computer, and I began surfing first gopherspace, then as it emerged, the Web. Since then I have gone much further—I'm now the technical lead on a project to edit all existing Chaucer criticism in electronic form, I create web pages on a regular basis, and I spend more time in front of the computer than I ever did in the library. Today I got 68 e-mails, from people all over the world, people I collaborate with even though we've never met physically. When I go to professional meetings, I find myself saying to new people, "Oh, I recognize your name from e-mail . . ." and they say similar things to me. Thanks to computer connections I have friends in many places who look things up for me (not just literary files, either—the Japanese are great for bootleg rock CDs, for instance . . .) The computer is a phatic connection to writers I sometimes never see. It makes me a less isolated writer and reader.

But only in the last four or five years have I begun to really realize how these kinds of collaborations can transfer into the writing classroom, how I can use the computer as a composing space, and not just a typewriter on steroids. In a way, it's been my students who have taught me this—as they forward MP3s or virus warnings, tell me about how they interact through instant messaging, cut and paste passages for my comments. Cyberspace is a place to write—it doesn't matter where the writers are physically located. It is a composing space. As more and more of my students get comfortable composing on the keyboard, I have shifted more and more time to teaching writing here, even though we don't have networked computer classrooms. The computer hasn't replaced f2f teaching for me, and certainly not f2f conferencing, but it has made it easier for me to do more teaching and more conferencing and different kinds of both with my students. It has created a new kind of community for us, one where we are connected both phatically and pedagogically. The more I study it, the more I like it.

Steven D. Krause

What's the most important aspect of the computers and writing community for you, and why is it so important?

The "tribal" aspect of the computers and writing community is important to me, of course. Online and face-to-face at conferences, we are friends and comrades who seek each other and find each other because of the commonality of our experiences. But the thing that's most important to me about the C&W community is its role as a space for the exchange of vibrant ideas that are so useful to me in my teaching and my life. Both online and at conferences, people in the C&W community ask and get answers to fairly practical "how to" questions (everything from "how do you incorporate MOO discussions in first year research writing courses?" to "how do you turn off the auto-correct features on MS *Word?*"), and more theoretical "why" questions ("why should I use computers in my writing classes in the first place?" and "why are completely online classes a good or bad idea?"). I can think of dozens of times where I've benefited either from my attempt at giving a version of an answer to someone, or where I've had a question that was quickly answered by many others. In theory, all academic communities are supposed to be like this, but in practice, I've found the computers and writing community to be one of the best representations of this ideal.

Deena Larsen

Why do you choose to be active in the computers and writing community?

All my life I have wanted to write in a way that just doesn't work on paper. I started out with stories that you can read forwards or backwards, poems that mean something if read across the page and something else if read down the page. On paper, you have to explain how to read this (to follow this thread, go to page 76). On a computer, it is as easy as point and click. I wrote a large work (over 100 poems) about women in a small Colorado town in the 1800s, and put the piece on a model railroad bed, with strings showing the connections. Without a computer, only a few people could see this work. On a computer, I shared and expanded my concepts with graphics, maps, connections, hidden passageways, and links. (This work, *Marble Springs,* is available on disk from Eastgate Systems.)

Using a computer, you can add new dimensions and create new forms of literature. You can add sound and motion. You can set up a reading so readers can choose to read in different orders. You can show the structure of a work and readers can click anywhere in the structure to see the underlying plot and meaning. In little over a decade, a new writing community has evolved to explore the infinite array of possibilities in computers and writing. I have a variety of works on the web that use everything from ferris wheels to the periodic table of elements as structural analogies for my stories and poems. Electronic literature, new media literature, hypertext: whatever term you use, there is now a new genre of writing that can be done only by using a computer. The Electronic Literature Organization (ELO) <http://www.eliterature.org> has a directory of these works.

Computers also bring far-flung writers together. Although there are only a handful of electronic writers in Colorado, I can meet with my fellow writers without traveling long distances—online. For the past two years, I have hosted chats for ELO where electronic writers come together to discuss how works on a computer differ from works on paper and the new directions we can follow for these works. Writers workshops, bulletin boards, courses, and other online discussion forums help us keep up with the latest technologies and avenues to exploit for our writing. For me, computers allow me to write in ways I want to write, and to communicate with others who are using the computers in ways paper can't even imagine.

Lisa Lebduska

What scholarly project in computers and writing has been most influential for you, and why has it been so influential?

The Computers in Writing Intensive Classrooms at Michigan Tech University (Summer 2001), a vibrant example of humanist technology, has been the most influential scholarly project on my professional and personal development. Two weeks among knowledgeable, thinking and communicative technologists shone a gentle light on my pedagogy and activism within technological culture. Readings and discussions deepened my scholarly background, while hands-on activities and general policies insisting that lab assistants remain all talk and no appropriation enhanced my technological courage. A concluding exercise in which colleague Teena Carnegie introduced me to MOOs, for example, led me to incorporate a MOO into a sophomore-level writing course—a unfathomable leap for me prior to CIWIC. Overall, the experience provided an artful and sustaining blend of knowledge, challenge and camaraderie that I will endeavor to bring into the classroom and the scholarly community at large.

Paul LeBlanc

What's the most important aspect of the computers and writing community for you, and why is it so important?

While many areas within the Humanities cast a broad net in terms of inter-disciplinary work, no area compares with Computers and Writing in terms of the breadth of interests, range of activities, and types of people it embraces. C&W is an area where graduate students often play key leadership roles, where practitioner-teachers stand shoulder to shoulder with researchers, where the activities at the community college and the research university and the small liberal arts college are equally valued, where people move in and out of academia, and where the connections to other disciplines and pursuits is not only welcome, but actively pursued. It is one of the most varied and rich areas of academic study.

In that broad and generous spirit, I have particularly valued the way the established members of the community, often graduate faculty, have mentored and supported newcomers, whether they be graduate students or veteran faculty just discovering this field. Charlie Moran offered a gentle guiding hand to me when I was starting out and I have seen that same impulse from people like Cindy Selfe, Gail Hawisher, Fred Kemp, Bill Wresch, and many others. The word "family" is grossly overused in academia ("the department is like a family"; "the college is one big happy family" and so on), but the C&W community has often felt that way to me.

Bruce Leland

Why do you choose to be active in the computers and writing community?

I have been sustained throughout my teaching career by new challenges. At first it was new courses or new approaches to try out. When I started teaching with technology and eventually teaching about technology, I discovered a field that would provide continual challenges and thus would continually re-energize me. In this world there is no standing still.

I often learn of the new technologies from my students. More often, I learn from the computers and writing community—not just about the technology, but also about how it might be used and ultimately what it means. Whether on listserv lists, in MOO meetings, or at conferences, the unending conversation provides me with new ideas and makes me clarify what I've been doing (and what I want to do). The people who teach with technology, teach about technology, and theorize about technology are a diverse, generous, stimulating, and visionary group of colleagues. I'm active in the computers and writing community because the community is itself so extraordinarily active.

Bernadette Longo

What scholarly project in computers and writing has been most influential for you, and why has it been so influential?

My current scholarly project looks at relationships between language that was used to introduce computers into popular culture and notions of (im)possible human–computer relations that grew up around that language. This research influences my thinking about how we (in the U.S.) conceive of our relationships to the computers with which we write and teach writing. If we in part think of computers as machines that store our work in their memories, how are our human memories affected? If we fear that computers are actually more robust and intelligent than humans, what actions do we take to control our creations? This exploration of computers as human creations that mirror our humanity leads me to question how we work with these machines in classrooms. I look to younger students to help me understand these relationships, because they have grown up alongside computers in ways that were unknown when I was a child.

Jane Love

How did you come to be active in the computers and writing community?

In August of 1994, I had just completed my PhD in English at the University of Florida with a theoretical dissertation on the ethics of reading. The theory market being what it was, I accepted a postdoc position at UF that fall and prepared to teach my usual composition and lit classes. The English Department had other plans, however. IBM had just endowed UF with one million dollars to establish something that sparked the envy (and disbelief) of the engineers and scientists on campus: a networked writing environment.

Computers and writing was new to UF, and the task of figuring out a pedagogy equal to the Networked Writing Environment fell by default to a group of enthusiastic graduate students, and, alas, to me. My computer skills at the time consisted of word processing. Period. I had never used a modem, never heard of the World Wide Web, never sent or received an email, never heard of MOO or telnet, didn't know what a browser or a shell was, and thought that hypertext was the Macintosh HyperCard application. When Anthony Rue told us that the Networked Writing Environment would be running on a Unix server, I wondered why no one else in the room seemed disconcerted that our system was named "Eunuchs."

It was an inauspicious beginning. Within a couple of weeks, though, I had dipped into MOO and succumbed at once. I logged hours and hours in UF's MOOville, intoxicated by the sheer pleasure of immersion in language (this was before Web-based clients; my first logins were made with raw telnet, and then, soon after, with Tinyfugue). In the MOO, I could play with theory. The conceptuality of language was so . . . *tangible* in the MOO. It was like leaving a sanctuary and walking into a warm and sunny kitchen, where through cooking and food and imagination and love, the spirit is viscerally made flesh. For me, the MOO was pure *poiesis*. The problem of pedagogy evaporated. I began to teach, and eventually to write, in the MOO.

Because I had no formal background in composition and rhetoric and only dubious technical skills, I was at first intimidated by the computers and writing community. (The knowledge, creativity, and expertise of the people here still humble me.) But this community is axiomatically pragmatic and welcoming: it's not where you come from, it's what you do, or want to do, that matters. The tastes for risk, experimentation, and vision run strong here; they make this community a good theory kitchen. Everything I know about collaboration, I've learned here. And it's not true that too many cooks spoil the soup: they make it richer, more complex. They make more soup. :-)

Marjorie Coverley Luesebrink

What's the most important aspect of the computers and writing community for you, and why is it so important?

The Computers and Writing community serves several important functions in the larger network of academic/artistic activities. Because it links the educational effort with critical work and many kinds of experimental literature and art, it provides a valuable bridge for both practitioners, scholars, and instructors to explore new frontiers in both creating for and teaching with electronic communication strategies. Also, the Computers and Writing community is actively engaged in defining the stunning re-mediation of text and image in this new medium. Investigations into online collaborations, the narratives of MOO/MUD, the "literature" of discussion groups, and the language of hypermedia stories and poetry have contributed to our understanding of how we might analyze and teach writing using new kinds of texts and pedagogy.

In addition to providing a forum for investigation of teaching and writing in new media, the Computers and Writing groups have been very active in expanding the kinds of resources that teachers can use as content material. Online libraries, literature, and e-literature can enrich the reading/writing experience for students even as the computer is improving access to instructional frameworks. By encouraging teachers to use hypermedia literature in their reading lists, for example, the community extends and reinforces the idea that electronic syntax (both literary and expository) and techniques can be studied, analyzed, and expanded.

Karen J. Lunsford

What worries you about the computers and writing community, and why does it worry you?

What worries me about the computers and writing community? The name—or rather, how some people from other areas of or outside of my university react to the name. It has the same problem as "composition" or "writing course." People in the know recognize that both "computers" and "writing" may be defined to include a broad range of technologies, issues, methods and interests. In fact, some people who identify C&W as a "community" may even wish to limit the definitions somewhat in order to provide a more central identity. (I don't think that's the way to go.) However, to people who are not familiar with C&W scholarship and conferences and MOOs and listservs and networks and websites and programs, the words "computers and writing" connote much more restricted and often negative meanings, including the idea the C&W teachers merely show students how to use word processors and how to apply grammar/usage rules. Even open-minded colleagues have initially responded, "I don't teach that," when I have asked them to submit proposals to C&W conferences. Ironically, they do teach "that": computer-mediated communication, Internet studies, user studies, user documentation, participatory web design, database and interface construction, electronic journals, multimedia studies, and so on. When they have attended the C&W conferences, they have been delighted to find perceptive audiences interested in their work. Still, I am concerned that because people who identify with other disciplines do not quite realize what it is we do, much C&W scholarship is under-cited in texts that would benefit from our work. And this goes the other way as well: I would like to see more C&W people at conferences such as the AoIR (Association of Internet Researchers).

Barry Maid

How did you come to be active in the computers and writing community?

I started using computers to teach writing in the early '80s. While by default I ended up as the person who "took care of the machines," I wasn't the department's "computer person." I was too busy being WPA and Department Chair. We had hired Andrea Herrmann, who was one of the pioneering people in word processing research. Through Andy, I met Cindy Selfe, Gail Hawisher, and the others. In fact, I even did a review for *Computers and Composition* sometime in the mid-'80s. Then, as I moved out of admin work in the early '90s, I had time to explore the emerging possibilities on the net. I found MBU, Netoric, and other places. Through my participation in these nascent virtual communities, I met Jennifer Jordan-Henley, and the Writing Center Consultation Project was born. I guess the rest, as they say, is virtually history.

Nick Mauriello

How did you come to be active in the computers and writing community?

In 1993, I was an adjunct instructor in the English department at a university in New Jersey. The English department was under pressure to integrate computer technologies into the curriculum from the College of Business, which was seeking national accreditation. Being the youngest member of the English faculty, the task was pushed down the totem pole to me. My department simply assumed that since I was young, I knew how to operate a computer. Well, like most of us in the field today, when I started I was a part-time employee who knew very little. I was a recent Master's Degree graduate fresh from a summer studying Shakespeare at Oxford. Yet, my credentials seemed fine for the job!

To make a long story shorter, what I witnessed that semester was amazing. The students became engaged in their writing assignments and collaborated freely. They helped each other with the technology. They revised more and their writing improved. All of this from Novel's *WordPerfect* 4.1!

This one semester led me to consider a PhD in computers and writing, which I started in fall 1995, at Indiana University of Pennsylvania (IUP). During my first semester I met Gian Pagnucci who was a new tenure track assistant professor. To no one's surprise, Gian was asked to develop the English department's computer technologies. Gian and I became fast friends. We developed composition courses using the then new killer application, the graphical Internet browser. As we collaborated we became somewhat isolated from our colleagues. Most just didn't understand what we were doing. First, we didn't teach the highbrow literary theory courses—we taught composition, a blue-collar art. Second, many of our writing colleagues resisted the initial shift to curricular technology, marginalizing us further within our department. It was this perceived sense of isolation that led us to the computers and writing community.

As Gian and I sought out the active members in the computers and writing community we found others who encouraged us to continue with our classroom research. Lisa Gerrard, from UCLA, was editing a special issue for *Computers and Composition* and accepted our first article for publication. Lisa then introduced us to Deb Holdstein who edited an early computers and writing book with Cindy Selfe. As we became friends with Deb and other experts in the field, we decided to hold a small symposium on the future of computers and writing. In the fall of 1998, twenty-eight scholars answered our invitation. Even now, the list seems incredible: Kathleen Yancey, Michael Spooner, Myron Tuman, Christina Haas, Lee Odell, Bill Condon, Helen Schwartz, and Joe Trimmer to name a few.

What Gian and I found was a receptive and active computers and writing community, one that continues to fuel our research. As Gian and I finish our third book in as many years, we are keenly aware that our success is indebted to those computers and writing scholars who welcomed us into the community.

Leland McCleary

How did you come to be active in the computers and writing community?

In the early 1990s I was participating in a number of LISTSERV groups that focused on second language teaching and applied linguistics. One of these groups was SLART-L (Second Language Acquisition Research and Teaching), and it was in that group that I learned of MBU-L (Megabyte University), to which I subscribed in October of 1992. I was immediately attracted by the tone of the discussion, which combined incisive and beautifully reasoned argumentation with the friendly banter of colleagues who have become friends. Since I had gotten into the habit of saving list mail, I started archiving the discussions. Those first fourteen months of MBU-L became the corpus for my doctoral dissertation on various aspects of discussion group discourse. I couldn't have chosen a better corpus, because I had to read and reread those messages hundreds of times, and I never tired of it. That was my initiation into Computers and Writing: I was privileged to join an ongoing discussion among the people who had most thought about it and were most involved in exploring it.

At first I felt like a guest of the conversation, since for twenty years I had been outside the world of English departments and First-Year Composition, where keywords like "rhetoric" are very different from what you hear in TESL; but at the same time I felt a tremendous affinity. That affinity remained virtual until 1993, when I was able to spend two months with John Slatin at the Computer Writing and Research Lab in Austin. Then in 2000 I was able to attend my first Computers and Writing Conference, in Fort Worth. It felt like a homecoming, and I was meeting most of the people for the first time.

Dan Melzer

What's the most important aspect of the computers and writing community for you, and why is it so important?

I think the most important aspect of the computers and writing community is its breadth: the way it influences everything from classroom pedagogy to writing center work to Writing Across the Curriculum. So many of our writing classrooms are now "computer classrooms" that it's becoming difficult to see "computers" and "writing" as two separate entities. Anyone interested in the future of writing center work is also interested in Online Writing Labs, and discussions of OWLs and email tutoring are as frequent on listservs like wcenter as discussions of distance learning are on WPA listserv and discussions of Electronic Communication Across the Curriculum are on WAC-L listserv. Recently at CCCCs panels on computers and writing have grown exponentially, and almost everyone working in the field of rhetoric and composition is touched by computers and writing in some way. So even writing teachers who don't consider themselves part of the computers and writing community are still influenced by this community. Not too long from now, we may simply think of the writing community as the computers and writing community.

Barbara Monroe

How did you come to be active in the computers and writing community?

Here is my conversion narrative, with John Slatin as the messiah figure. It was a quiet summer day in 1991 in a deserted computer classroom at the University of Texas facility when John Slatin gave me a personal introduction to the Daedalus program. Within five minutes, I knew that I was looking at something that would truly support teaching practices that I had only imperfectly approximated in my classroom without computers. Put another way, I instantly and intuitively knew that I would no longer have to stand at a photocopy machine, making copies of student papers for distribution and peer critique in class. That day also marked the beginning of a new career orientation for me—from American literature to composition and rhetoric—because I was drawn to a community of scholars who valued teaching, a community where professionals also practiced what they preached, collaborating and supporting each other's work, in far less competitive way than I had seen in other areas of English studies.

Looking at my conversion narrative now, I see important principles that hold true for introducing "new adopters" to instructional technology. These principles are especially important to me now, as I work with future English teachers who seem reluctant to adopt communications technology, much less online learning environments, as part of their teaching repertoire. In my own story, notice that someone sat down with me and introduced me to the software, discussing with me not only the how-tos but also the why-tos. And notice, too, that I quickly adopted the software because it confirmed—not challenged—my own beliefs and teaching practices.

That's what worries me about the future of the computers and writing community. As others have noted, our numbers have remained relatively stable over the years. Our community has not grown in numbers. Unlike what others have said, however, I don't believe the field has been subsumed, that we have "won." What we need to realize as a community, especially if we hope to expand our influence in public schools, is that certain values are deeply embedded in the architecture of school life as well in the teaching practices that that institutional structure supports. If internetworked computers are to transform schooling as we have known it for the past 150 years, we need to talk first about teaching values, not hardware and software, and then allow the conversation—and the community—to go/grow from there.

Virginia Montecino

What scholarly project in computers and writing has been most influential for you, and why has it been so influential?

My scholarly interest in computers and writing centers around Internet Studies, Cyberculture, how the nature and form of digital information is constantly evolving, and how students' writing experiences can be enriched by publishing in a Web environment. It is impossible to discuss literacy in this digital age without critiquing the medium of the Internet as a communications tool and as a social, political, professional, and cultural phenomenon.

My journey with computers and writing began in the mid-1980s, in the days of old 8088 PCs with no hard drives, running DOS (no Windows). The programs had to be used from floppy disks. Since then I can't imagine teaching and learning without incorporating computers and the Internet into the pedagogy.

I have developed an extensive Web site <http://mason.gmu.edu/~montecin/> for teaching and learning using computer technologies, with special emphasis on how to use the Internet for research and writing and critiquing the Internet as an integral part of the new literacy in this computer age. The Web site grew out of necessity when I was teaching my first distance learning advanced composition class and I needed a vehicle to distribute material quickly and easily to my students. Since then my students also publish a significant amount of their academic writing on their course Web sites. I don't consider my students to be truly accomplished writers until they learn to compose for and operate within a wired environment.

Students who don't learn to write and publish in a Web environment will be at a disadvantage in this wired world. Composing for a Web environment is not simply a matter of publishing linear texts using html. Faculty and students have to be cognizant of and open to writing in new forms, using new organizational strategies, more integration of various elements (such as graphics, sound, video) and rethinking the meaning of audience.

Michael R. Moore

How did you come to be active in the computers and writing community?

In 1997 I was working with a small, informal working group at the University of Arizona on computer-mediated pedagogy issues. We were trying to figure out ways to incorporate various writing and design technologies in the English curriculum. By 1997, of course, people in various fields had already begun to question many of the acritical, overly enthusiastic claims made about computer-mediated pedagogy, and our group was interested in designing and assessing curricula that allowed for critical and inquiry-based approaches to teaching, learning, and researching in technology-rich environments. As anyone who's been through this exciting process knows, we soon found ourselves knee deep in both historical and contemporary scholarship in intellectual property, multimedia and hypertext theory, new approaches to ethnography and collaboration, literacy theory, modes of consumption and production, and equitable access to both technology and to education. For many of us, these collaborative inquiries added (and continue to add) new and reinvigorated approaches in our writing classes.

Very exciting times.

As part of our explorations, we discovered several online and in-person communities who were doing similar kinds of work. Some of them were disciplinary in nature and emphasis—art; communication; anthropology; literature; economics; political science—and some were broadly, robustly, and mind-bogglingly interdisciplinary, such as the computers and writing community. We soon discovered and met folks who were combining research and scholarship in first-year composition with computer-programming languages; technical communication with multimedia applications; service learning with hypertext and desktop-publishing emphases. Many of us began—and continue—to present our work at the yearly Computers & Writing Conference, where we continue to meet new colleagues and collaborators from across a broad range of interests and disciplines. An important lesson for me has been that, while working with technology in a liberal-arts or humanities context can sometimes be an isolating experience, it does not have to be: both our local working group and the larger computers and writing community provide genuine, sustainable, and necessary support for our work.

Charles Moran

What scholarly project in computers and writing has been most influential for you, and why has it been so influential?

The scholarly project that I've found draws me most in our field is the intersection between technology and socio-economic class. An important early work, C. Paul Olson's chapter, "Who Computes?" made important points that have pretty thoroughly disappeared from our discourse. In 1987 Olson clearly saw that new technologies were affordable to some, not affordable to others. Further, he saw that these technologies advantaged those who had access to it, and, conversely, disadvantaged those who did not have access to it. In my own book chapter, "Access: The A-Word in Technology Studies?" I drew heavily on Olson for inspiration and direction, really updating his chapter for the second millennium. Cynthia Selfe and I took the argument further, applying it to American K–12 education in our 1999 article in *English Journal.*

This project has led me to look at ways in which new technologies are affecting American postsecondary education. Buying and maintaining thousands of computers costs a huge amount of money. Where do institutions find this money? By raising tuitions (thereby excluding those who can't pay) and by reducing staff (thereby increasing student/faculty ratios). If postsecondary education were a business, this would seem fine. But it's not: education is people-intensive. And so we find ourselves needing computers and needing people, and needing to pay for both. Wealthy institutions can do this; poor institutions can't.

Technology has done here what it so often does: make what was before transparent suddenly visible. There's always been a wealth gap between individuals, between colleges/universities. Technology has made this gap suddenly and powerfully visible. It makes us see that the same people who don't have access to technology lack access to food, housing, good legal advice, health care—indeed, to anything that money can buy. I hope that we have the fortitude to take what we see and to translate our new knowledge into action.

Paul "Skip" Morris

What worries you about the computers and writing community, and why does it worry you?

The computers and writing community is made up mostly of computer enthusiasts like myself. We publish articles and books that are read, in the main, by members of our community, and we attend conferences that appeal to the teachers (our community members, once again) who are interested in what computers can do in the classroom. However, many teachers see no genuine relationship between literacy and computer technology in the way Cynthia Selfe talks about their connection in her article "Technology and Literacy: A Story about the Perils of Not Paying Attention." For example, my department colleagues only want to know enough about computers to get the job done. The workshops I give on computers and writing are sparsely attended, probably because many teachers see computers as glorified typewriters. I'm worried that we are not convincing our colleagues that computer technology enhances literacy learning.

Joe Moxley

What's the most important aspect of the computers and writing community for you, and why is it so important?

I've taught in computer classrooms since 1984. Throughout this time, I've been interested in research in *Computers and Composition,* and I've enjoyed attending a number of C&W Conferences.

As a writing teacher, I've been reinvigorated by new communication technologies. My current students' writing projects are dramatically different from the works of my students back in the 1980s. I mean, I still remember when I was impressed with color bar charts. But now my students are developing interactive Web sites, creating three-dimensional models and animations, and integrating video. I enjoy working with my students in computer classrooms, focusing on projects that have real audiences and implications. I feel that email and working on class ezines, such as http://toolsforwriters.com, have brought me closer to my students—and made my work more meaningful.

For me, electronic theses and dissertations are the most important aspect of the computers and writing community. The NDLTD (Networked Digital Library of Theses and Dissertations) is an extraordinarily interdisciplinary and international group. . . . I'm really excited about helping develop a new genre of scholarship, as suggested by http://etdguide.org/ETDList.asp.

After the Internet, it's difficult to imagine "What's Next?" But I suspect colleagues in the C&W community, particularly graduate students, will be leading the way.

4

Integrated Meaning-Making Systems in Computers and Writing: Cyborg Literacy

In the cyborg era, one of the most important conversations in the computers and writing community has been about the way meaning is made across contexts and with various technologies, especially the computer. Among prominent questions asked specifically about meaning-making and technologies are the following: Is there something unique about the way we read, write, and speak with technologies like the computer at the same time there is something unique about the various contexts in which we operate with them? Do we make meaning in new and different ways, or are we simply seeing a broadening or adaptation of what has come before? What specific implications does our meaning-making with technologies like the computer have for the evolution of meaning-making itself and for us as individuals?

The term *meaning-making* emerges from semiotics and operates somewhat interchangeably with *literacy,* if literacy is conceived broadly and not tied only to alphabetic characters and structures (Hodge & Kress, 1988). Specifically, meaning-making requires that symbols and signs must be understood as the building blocks of meaning. In this way, almost anything could be understood as having meaning-making capital, if it is interpreted somehow by someone. If I were to purchase a McDonald's Extra Value Meal™, for instance, then it's not simply my reading of the menu and placing an order, paired with the employee's interpretation and completion of that order that's interesting, although those interpretive transactions certainly would be interesting. Also, it's what that meal itself represents, as both the employee and I construct it. My choice of foods may reflect my

159

cultural identity and background, for instance, or it may emerge from a specific social context where a specific order was privileged or otherwise emphasized. I might live in a region where McDonald's is running a particularly aggressive marketing campaign for one of its meals or where it is the only restaurant to serve a particular meal. At the same time I describe the breadth of meaning-making as a concept, I should stress that it does not de-emphasize the role of alphabetic characters and structures; it simply makes them one of a multitude of bases for meaning-making, rather than the only one.

My contribution to the ongoing scholarly conversation in the computers and writing community is to introduce a new term for contemporary meaning-making suggested by the cyborg era now before us: *cyborg literacy*. I define the term in interdisciplinary fashion, choosing ultimately to represent it as the integration of a series of systems that compel simultaneous attention to individuals, technologies, and other elements in the contexts they share. Finally, I present a multidimensional case study of cyborg literacy, demonstrating how it provides a unique view of meaning-making today.

DEFINING CYBORG LITERACY

A number of terms for contemporary meaning-making have been introduced: *technology literacy* (Hayden, 1989; Lewis & Gagel, 1992; Selfe, 1999a; Thomas & Knezek, 1998; U.S. Department of Education, 1996), *computer literacy* (Bitter, 1986; Lombardi, 1983; Luehrmann, 1984; Masat, 1981), *media literacy* (Bruner, 1999; Potter, 2001; Silverblatt, 1995, 1997; Tyner, 1998), *information literacy* (Arp, 1990; Association of College and Research Libraries, 2000; Behrens, 1994; Bruce, 1997; Rader, 1995; Snavely & Cooper, 1997), *digital literacy* (Gilster, 1997; Lester, 1995), *Internet literacy* (Hofstetter, 2000), *electronic literacies* (Sullivan & Dautermann, 1996; Warschauer, 1999), *multimedia literacy* (Hofstetter, 2001), and more. These cannot be taken as equivalent views, however; they represent different values and perspectives. Computer literacy, for instance, is often defined as a skills set; to be computer literate, that is, means being able to perform successfully a finite number of skills, like typing a paragraph on a word processor or saving a file to a floppy disk or the hard drive. Information literacy more often references an individual's ability to interact successfully with information structures, like library databases and online references. Different still, technology literacy is typically defined as the ability to use multiple technologies like the computer and think across them; the definition is less about the skills anyone can perform with the technologies and more about a general familiarity with their operation. With any of these terms, the use of the word "literacy" implies a connection with meaning-making in the print tradition, rather than a separation. For scholars who have introduced the various terms, the distinguishing feature of their work has been the word or words paired with literacy.

Another prominent view has been that what we are seeing is a transitioning of literacy itself, rather than new forms. Indeed this view has been perhaps the most popular, spanning from early works like *English and Reading in a Changing World* (Evertts, 1972) to more contemporary examinations like *Page to Screen: Taking Literacy Into the Electronic Era* (Snyder, 1998). Scholars thinking about literacy as changing to include attention to meaning-making with computers and other technologies generally believe that the technologies do not reflect a fundamental transition, like the more often recognized shift from orality to literacy. That is, they do not offer anything distinct and sophisticated enough to signal a meaning-making transition of that magnitude, and they serve like any other technologies associated with meaning-making. In "The Disappearance of Technology: Toward an Ecological Model of Literacy," Bertram C. Bruce and Maureen P. Hogan (1998) argue, for instance, that any technologies become naturalized into meaning-making contexts in which they are often employed. They write

> As technologies embed themselves in everyday discourse and activity, a curious thing happens. The more we look, the more they slip into the background. Despite our attention, we lose sight of the way they give shape to our daily lives. This disappearance effect is evident when we consider whether a technology empowers people to do things that would be difficult, or even impossible otherwise. (p. 270)

Some scholars, like Sven Birkerts (1994), author of *The Gutenberg Elegies,* suggest that computers and other technologies are bringing problematic changes to literacy, thus that we should not spend much time with them. The perspective in such cases is that literacy is being somehow corrupted or defaced by the technologies.

Despite the new terms scholars have introduced and the alternate arguments scholars have forwarded about the potentially changing shape of literacy itself, it's fair to say that we do not yet have a full and complete understanding of contemporary meaning-making. Not everyone believes that specific references to literacy are appropriate for such terms, even. Anne Wysocki and Johndan Johnson-Eilola (1999) write, for instance, "Too easily does 'literacy' slip off our tongues, we think, and get put next to other terms. . . . Too much is hidden by 'literacy' . . . too much packed into those letters—too much we are wrong to bring with us implicitly or not" (p. 349). Indeed terms without literacy have been suggested. Gregory Ulmer's *electracy* (1998) is among the most prominent; he means it to represent the state of meaning-making at the time that its central apparatus is electronic media. I am not sure that we can ever reach electracy, and scholars have certainly problematized apparatus theory as being too essentialist. However, electracy does provide an important possible trajectory for meaning-making. As Walter Ong (1982) traced a path between orality and literacy, so we can think about contemporary meaning-making as between literacy and electracy, whether we believe electracy to be possible and reasonable or not. Such a path enables us to locate possible terms for meaning-making more carefully, positioning them between two constants and crafting definitions for them to reflect that positioning.

Despite Wysocki's and Johnson-Eilola's caution and Ulmer's electracy, I believe contemporary terms must include literacy because of the connection it enables to the past and the present. Here I am arguing for a connection to literacy in new terms, rather than the sense that literacy alone still proves appropriate. Literacy scholars like David Schafsmaa (1993) and J. Elspeth Stuckey (1991) have shown us, for example, that meaning-making sometimes involves social and cultural violence, inequity, and injustice, as often as confidence and success. This charged reality parallels how technologies like the computers have brought their own problematic implications, such as the way affluence often ties to access and the changing but still much too narrow state of diversity in networked spaces. If terms representing contemporary meaning-making include literacy, then they can benefit from both bases of knowledge, not just that specifically relating to technologies like computers. More, referencing literacy in new terms invites broad participation in conversations about them. That is, scholars who know a great deal about literacy, but not technologies, can share rich and challenging conversations with those who know such technologies well, and both parties can grow intellectually in important ways, as well as potentially advance knowledge about meaning-making in general.

With this rationale for including literacy in new terms, the question becomes what word or words should be paired with it. As I described earlier, a number of possibilities have been proposed. Interestingly, though, these terms rarely if ever focus on the individuals, technologies, and other elements of contexts they share simultaneously in meaning-making transactions, choosing instead to emphasize only the technologies at work, the media enabled by these technologies, or the information shared through them. The only term for contemporary meaning-making thus far to include individuals, technologies, and other elements in their shared contexts directly and explicitly is William Covino's cyberpunk literacy (1998), which he defines relative to Case, the central figure in William Gibson's cyberpunk novel *Neuromancer:*

> Ostensibly, *Neuromancer* narrates the future of cyberpunk literacy, with the portrayal of . . . Case . . . as a talented cyberpunk who can navigate and invade virtual and material realities from his computer console. Thus the novel maintains advanced technoliteracy—the cyberpunk riding the matrix—as a version of magic; as such cyberpunk literacy operates as a transcifiormative and transcendental force associated with the control and manipulation of reality. (pp. 34–35)

Covino's emphasis is certainly on the magic, that force that shapes reality. However, the means by which he characterizes cyberpunk literacy is Case, as the character rides the matrix as an increasingly self-aware agent in cyberspace. In this way, Covino's work proves important, offering a model for how to keep the emphasis in constructions of meaning-making on individuals, technologies, and other elements in the contexts they share. The question becomes, though: How can we adapt this emphasis to explore our meaning-making, if we are not ourselves cyberpunks, like Case, riding matrices in our real and virtual worlds and if we're not certain how much magic is really in our midst? What happens, I mean, when we need a theoretical model that can be located and evaluated in the computers and writing community's practices?

Cyborg literacy, this chapter's term, offers an answer by keeping simultaneous emphasis on individuals, technologies, and other elements in their shared contexts and offering a more observable and even measurable approach to understanding contemporary meaning-making. I propose we think about cyborg literacy in terms of systems—complex real and virtual systems. Such systems are multidimensional and broadly defined, as my use of cyborg implies; most important about each is what individuals, technologies, and other elements in the contexts they share tell us about the dimensions of meaning-making. Thinking about meaning-making in systematic terms is itself not new, of course. Scholars like J. L. Lemke (1998) and Bonnie Nardi and Vicki O'Day (1999) have shown us, for instance, that ecological concepts have strong applicability for thinking about the complexities of meaning-making as they relate to advanced technologies. Lemke, in particular, makes a case for the concept of *ecosystems:* "Literacies cannot be adequately analyzed just as what individuals do. We must understand them as part of the larger system of practices that hold a society together. . . . If we think the word society means only people, then we need another term, one that, like ecosystem, includes the total environment" (p. 286). Lemke's approach is to talk about what he terms *metamedia literacy,* which describes meaning-making across media, so it's somewhat different than this chapter's approach, but the arguments are parallel: We need to know as much as possible about the contexts in which individuals interact with either media or technologies. I argue for technologies because I am most interested in what creates the specific systems, where Lemke is most interested in how individuals interact specifically in media enabled by the various technologies.

Cyborg literacy draws its definition ultimately from the integration of meaning-making systems, rather than one or two. It's important to suggest that these systems often will not have identifiable overlap; they are what scientists term *emergent,* meaning unique at every level of magnification and analysis. Consider a grassy field in Tampa, Florida, for instance. In this ecosystem, we might identify a number of organisms, like butterflies, snakes, and plants, like dandelions and sunflowers, and our approach would be to think about the system together, observing how the many elements interact. Now imagine if we magnified our view such that only a single blade of grass was visible. What we have is a much different ecosystem, one where we're likely to identify different elements. The same would be true, of course, is the view was perspective was broadened, rather than magnified. If we were looking across central Florida, we likely would not see individual elements of the original grassy field in Tampa; however, we would see many different system elements of interest. By thinking about the importance of multiple ecosystems for our definition of cyborg literacy, we have the ability to learn about a whole host of influences on meaning-making. At one level of magnification, we might explore a single work station environment, whereas another level may enable us to consider a classroom full of individuals and technologies. And the same is true for virtual contexts. At one level, we might examine the system around a single user and her or his associated personae, and at another, we could study a system featuring a host of users and technological components, like a virtual community.

As the integration of emergent meaning-making systems, cyborg literacy provides a much broader and more inclusive conceptualization of contemporary meaning-making than previous terms have been able to offer, especially as it enables a critical and careful examination of interactions between individuals, technologies, and other elements in contexts they share. At the same time, it does not represent a standard, and it does not have any easy to follow guidelines for definition. Cyborg literacy relies on the individuals studying meaning-making to make careful and reasonable choices about which systems prove important to be included, and it thus may be defined differently every time it is considered by different scholars. Cyborg literacy, it follows, offers scholars a broad license to study contemporary meaning-making; we can pursue whatever is interesting whenever it is interesting, as long as it features the integration of multiple systems in which individuals, technologies, and other elements in their shared contexts interact and as long as we are willing to explain our choices. All that is required is that we take professional and personal responsibility for the studies of cyborg literacy we construct and share.

CYBORG LITERACY IN ACTION:
A MULTI-DIMENSIONAL CASE STUDY

In the fall of 2001, I began a faculty position in the Department of English at the University of South Florida (USF), and the case study I present here emerges from that first semester. The two courses I was teaching both related to USF's program in professional and technical writing; they were titled, appropriately enough, Professional Writing and Technical Writing. In each course, I required print and electronic assignments alike, as well as oral presentations, and I required both individual and group work. Specific assignments in the classes included a software comparison matrix, a professional Web site, a collaborative business proposal, a documentation guide, and a usability test design and report. As defined in USF's curriculum, the difference between professional writing and technical writing is that the former includes business, technical, and scientific writing.

I taught both classes in a computer classroom, and the wide range of students' abilities and experiences is what started me thinking about cyborg literacy. In what ways, I wondered, do individuals, technologies, and other elements in their shared contexts converge meaningfully in the sorts of diverse meaning-making systems in which the students have been interacting? And how might thinking about the integration of specific systems in our class prove useful both as a way for me to learn more about the students and their meaning-making practices? I decided to do a 6-week research project to pursue the questions, and I began my study by sending the following e-mail to all students in both classes:

Hey, all—

A request here.

I'm interested in learning more about your experiences in this class and specifically the way you're operating in and around various technologies in our class for assignments and other class activities. What I'd like to do is make observations during class in the form of field notes and then schedule interviews with some of you to talk more about what I've observed.

Two questions, then:

- Will you agree to allow me to do observations in class? I will not make observations unless everyone in the class agrees.
- Would you be willing to be interviewed by me? These interviews would be conducted outside of class and will not be for extra credit or anything like that.

Some important things you should know:

- You do not have to say yes to either question, and whatever choice you make will have no effect on your individual assignment grades or on your final grade in the course.
- You can choose not to participate at any time. If you decide you'd rather I not make observations in class, I'll stop immediately. Likewise, if you decide that you do not wish to be interviewed, you will not be.
- I will be using what I learn from observations and interviews in a book I am writing about computers and writing. Specifically, what I learn in our class will be for a chapter about meaning-making, so I will be focusing on how you complete assignments, work individually, work with others, and more.

Thank you for considering this!

J. I.

In the e-mail, and indeed in all of the research I conducted, I was careful to indicate the relation of student participation to grading, as I wanted to be sure students would not feel compelled to agree to the observations and interviews. In the end, everyone did agree to the observations, but only seven students indicated a willingness to be interviewed.

After the agreement was reached, I began observing the class as a whole, seeking to learn more about the systems important to study. I continued my observations online, looking for related systems of note there. I now present the three emergent systems I believe most valuable for consideration in this case—two real systems and one virtual system—and I then integrate them into a table to illustrate the breadth and inclusiveness of cyborg literacy as a concept. In each system, I first identify individuals, technologies, and other elements in the contexts they share as a way of demonstrating the simultaneous importance of each in the system and the way each system is unique. Then I explore particularly important meaning-making practices in each system, working from observations I made as a participant–observer, as well as field notes I took in support of those observations and interviews with class members.

Real System One: Jose and a Workstation

I first focused on a student who I'll call Jose and his classroom workstation environment. I observed several technologies of note: the Dell computer Jose was using and its associated components, a cellular phone he had clipped to his waist, and a portable compact disc player with headphones that he was wearing while he worked. In the environment around Jose and the technologies visible, I made note of several interesting elements: a chair that seemed always to be set much too low, a smallish table almost completely covered by the computer and its components, three notebooks open to pages with writing on them, a diet soft drink, a nearly empty book bag, and sandals that he had slipped off after sitting down. As I will relate, observing and talking with Jose helped me to understand how all of these system elements converged.

I first wanted to learn as much as possible about Jose as an individual and his particular meaning-making practices, the sort of work many other scholars have typically done. In a larger system, such attention to an individual would not be possible, but because Jose was clearly prominent in this system, the attention seemed warranted. In our first conversation, after glancing down at his desktop, no doubt hoping I wouldn't notice the Instant Messenger window open, Jose indicated that his meaning-making experiences begin and end with his family, as he chats with his parents and three brothers online whenever he's using a computer, even in classes. Jose and his mother read together, for instance: "Yeah we're reading the Oprah books now. We IM [Instant Message] all the time. It beats paying money to the damned phone Nazis." Jose does acknowledge that his chat sessions don't help him to complete assignments and meet other deadlines: "I IM with friends all the time too. Sometimes like all day and I don't get anything done. Not a thing. And I'm mad, but not mad, you know? Anything else is straight dull." Among those dull experiences are "reading textbooks and crap like that." At the same time, Jose does note several school projects he liked, including the one on which we were working that day in class:

Project	Reasons for Favoring Project
Job application letter and resume	"All the time online . . . and this is real, like what we for technical writing class
Debate for political science class	"I can argue all the time . . . it makes class fun."
Observations for anthropology class	"I get to go sit in the bookstore all day and see what people do . . . and some of them are freaks. It's fun to pick them out."

What becomes apparent in Jose's list of projects is that he most favors social approaches to learning, and given his constant use and valuing of chats and other interactive technologies, that favoring makes sense. It's especially interesting in this case how what Jose values in both real space and virtual space seems shaped by his interactions and interest in virtual interaction.

In the system, I noted that Jose made regular and sometimes even constant use of the full range of technologies around him, not just the computer. Jose makes it a point to emphasize that he relies on more than simply a computer to make meaning, in fact:

> I've got to be plugged in, you know? I keep my phone on silent and I'm looking at it all the time in case someone calls. Sometimes I get e-mail on my phone too. . . . The music gives me energy. It's like if I don't have my music I can't write as good. I stare at the screen and I stare more and nothing happens. When I've got Jay-Z or Aaliyah or someone like that on I can jam anything for any class.

When it comes to using a computer to make meaning, then, Jose feels most confident when he's using other technologies simultaneously. In this way, we might expand the way *multitasking* is often defined to include not just using multiple applications on a computer at once, but also interacting with other technologies. Jose's use of his cellular phone to receive e-mail emphasizes this point; although e-mail remains largely computer-based, recent advances in technology have begun to bring about changes. And another element of Jose's use of technologies actually makes the same point. Right now, he uses a portable disc player to enjoy music, but advances in audio file sizing, especially the popular mp3 format, mean that we have many music-listening options available. Because Jose already relates his cellular phone and disc player to his own meaning-making practices, he teaches us that they play key roles, a lesson matching my observation of the system. Without these technologies, the implication goes, Jose might not be able to "jam," so it's the convergence of the technologies that most directly informs our developing sense of this system.

Jose and the various technologies described were not the only important elements in the workstation system, however. In Jose's and my discussions, for instance, he emphasized that the chair he had to use in the classroom also influenced his meaning-making practices because it made using the computer awkward and uncomfortable. Jose explained, his language demonstrating how frustrated he had become with the situation: "Can't they buy the right chairs? I think I'm getting that shit people get when their wrists hurt." Ironically, the chairs are easily adjustable, and I showed Jose how to raise the seat several times; in the end, though, he refused to make adjustments and instead believed the chairs were unchangeable and thus especially problematic. If Jose was indeed developing carpal tunnel syndrome, I worried, then that could have substantial implications for meaning-making in the system I was observing, plus his ability to make meaning in other environments. Not being able to type would effectively silence Jose in many computer-based class activities, after all, unless we could obtain and install speech recognition software. Even if that software was available, though, it would alter the means of meaning-making considerably by changing the rate at which Jose could respond in discussions and by limiting the rhetoric of the responses he might choose to make. It would be nearly impossible for Jose to use some variations of specialized discourse, like BRB (Be Right Back), ROTFL (Rolling On The Floor Laughing), and IMHO (In My Humble Opinion), for instance, unless he spelled them out or used the full words. In such a scenario, it is limited access to discourse in general that matters more than access to specific terms or phrases; not having access in general limits rhetorical choices.

In this system, simultaneous attention to Jose, technologies, and other elements in the contexts they share enables us to learn a great deal about meaning-making. We know that computers at workstations aren't the only technologies that matter, which shows why constructions of meaning-making like computer literacy that reference a specific technology or family of technologies fall short of the mark. We also learn that factors like an undersized chair can have significant implications. A construction of meaning-making not emphasizing the environment, along with the individual and associated technologies, cannot account for such important factors.

Real System Two: The Classroom

I next focused on the classroom environment. In this system, Jose was visible, but not distinct; each of his 19 classmates was also visible and important. If I include myself, 21 individuals were evident, and that number means that less attention can be given to specific individuals because what is most visible is the relationship between them. The classroom contained 35 desktop computers, including 27 Windows 2000 workstations and 8 Linux workstations, and 3 printers. Other technologies evident were 3 television sets, a VCR, a DVD player, 3 servers, 2 telephones, and a flatbed scanner. At times students brought in their own laptop computers and cellular phones, but not on the days I observed and took field notes. Primarily the table, workstations, and pathways between them defined the classroom environment. However, other elements were notable: a bookshelf with dictionaries and thesauri, two whiteboards, a chalkboard, and a small coffee table. It's also important to note that windows and often-inactive central air conditioning contributed to make the room extremely warm most days and that we also had an ant problem, despite being on the third floor of our building.

The students in the classroom system made meaning in a number of interesting ways, each valuable to examine in more detail. On the first day of class, I assigned groups that I intended to stay constant throughout the semester, but several of them had major conflicts in completing the first assignment, so I decided to introduce new groups. These new groups were completing their first activity together when I was studying the classroom system; for the activity, the students had to develop collaborative evaluations of professional communication software, like eHelp's *RoboHelp* and Adobe's *FrameMaker*. In each group, one or two students assumed leadership roles immediately—some naturally, others by being the loudest or most vocal—but the way the groups worked with respect to the classroom's computers showed a range of meaning-making practices. In Jose's group, for instance, a student I'll call Josh set the group's agenda and seated himself at a central computer, encouraging and in some ways requiring the others to form around him. As Jose put it, "We just say ideas and then he makes them work or say no." This group dynamic worried me, of course, in how Josh may be silencing others by being too directive and in control. Another group appeared to be led by a student I'll call Susan, and she showed a much more democratic leadership approach, making sure to involve everyone at all times. In that group, the members each used a computer, and they each located information that was incorporated into an assignment draft; then the group discussed the draft together, with Susan asking questions and listening as much as talking herself. What we can take from the student groups is that hierarchy emerges, whether encouraged and facilitated or not, and that seeing how groups interact with each other and around computers offers insight into the way they make meaning.

Studying the classroom system also showed that any history a meaning-making community has will continue to exert influence. Nowhere was this reality stronger than when the new individual groups presented their work to the class as a whole, as ill feelings from the first groups appeared to create a competitive atmosphere among new groups, even though their presentations were not for a grade. Students made it a point, for instance, to emphasize any information they found that was not shared by groups presenting before them and to do so almost strategically and aggressively, rather than imagining their contribution as helping the class as a whole learn as much as possible. Interestingly, Jose was one of the most prominent students in this regard; he began his comments to the class with "That first group, they, they [sic] like struck out. They didn't have nothing [sic] like we found." In the moment, this opening struck me as strange at best, but I hadn't yet begun to understand the competitive element at work in the classroom, so didn't know quite what to think. Later, though, when I realized that Sam, a student from Jose's original group who constantly questioned him in their discussions, was in the first group, Jose's comments made more sense. They were not any more appropriate, of course, but at least I knew more about their context.

Ultimately, while I wasn't naive enough to believe that changing the groups would make everything better in the classroom, I didn't anticipate the way past conflicts would still be evident and influential. We can learn from this experience that meaning-making in a group setting always involves any history that group shares, whether directly related or not, and thus that thinking about such history can help us make sense of group meaning-making.

From studying the classroom system, we learn a number of interesting and valuable lessons about meaning-making. In any system, hierarchies emerge, whether intentional or not, creating social, cultural, and political relationships among group members that may or may not support meaning-making. Additionally, any histories a system has will continue to exert influence on that system, even if changes are made.

Virtual System One: A Class E-list

The last system of note for this case study is the virtual community the class created via its electronic list, or e-list. Although I served officially as listowner and moderator, I set the list so that all posts would be accepted, rather than approve messages. In a virtual system like the community the class developed via its e-list, individuals, technologies, and other elements in the contexts they share require knowledge of both real and virtual influences, rather than just those virtual. That is, what we did in virtual space connected with what we did in real space, thus providing the relationship. Were the environment only virtual and home to individuals who had not developed relationships in real space as well, matters would perhaps be different.

In the virtual system, I needed to think about identity carefully because individuals' virtual presences represented a combination of real and virtual identities for everyone invested in at least one virtual and one real system associated with the class. On the class e-list, we did spend time defining ourselves specifically, as I required all of the students to collaborate with me in defining rules for participation and standards for participation, thus the way we should each represent ourselves through our posts and understanding of others' posts. I emphasized how I would sign e-mails J. I., for instance, rather than the formal Dr. Inman or the informal James, as a sort of middle ground between the formality that real systems often require and the informality that virtual systems often encourage. Some of the students took my choice as a model and signed e-mails with their initials, but others created virtual personas and explained in a post how their virtual selves emerged from their real selves. A student who I'll call Anita cast herself as "Starla":

> Im shy in class and teachers tell me I need to say more always. Starla isn't shy at all.
> When Im Starla I talk all the time. Sometimes I say things I wish I could take back.
> But that's what Starla is all about. She says whats on her mind even if it is mean.

I picked Starla because I heard [it] in a movie once. The lady was real loud and just said everything. Other people got angry but she didn't care.

Anita's Starla character online demonstrates what I mean about how thinking about virtual identity requires a convergence of real and virtual identities, in virtual systems emerging from real systems. Anita defined Starla by referencing herself and her shy tendencies in class, thus uniting the two. At the same time, she does see Starla as unique, and her contributions to the community reflect the success with which she played the role. In responding to a discussion thread about personal information in resumes, for instance, she wrote, "Im saying who I am and I don't care if they don't like it. Either hire me or don't." Clearly the brash comments Starla offered are not those that would be offered by Anita, so she contributed to the e-list system as Starla specifically. I didn't worry about those perspectives being reflected in Anita and Anita's work in the class, however, as I knew it was a role and that the role was helping Anita to express herself and learn more about meaning-making in the virtual system and specifically how such system participation can help students to develop as writers.

Thinking about technologies in a system enabled by technologies seems complex, but the adaptation required is actually pretty straightforward across both real and virtual environments. At the same time, I do believe that many researchers do not explore the technologies apparent in virtual systems. In the e-list system I'm describing now, for instance, it's significant that it was hosted by Yahoogroups because other hosts might require a different interface for users, whether those who access messages via the World Wide Web or instead those who receive e-mail directly from e-lists. Yahoogroups works much like popular Lyris e-lists in that a Web interface enables users to subscribe, unsubscribe, change settings, and read messages. Other technologies prove much different, however, and listserv, listproc, and majordomo are examples of note in this regard. Where an individual can subscribe to Lyris lists by completing an online form and replying to a confirmation message, she or he would have to send e-mail directly to listserv, listproc, or majordomo systems; these technologies do not themselves make a Web interface possible, though certainly technologists have developed scripts that enable Web subscription by automating e-mail requests. In the e-list system, where the technologies at work become most visible is in the many different ways students manage their subscriptions and subsequently how these management approaches determine in some measure their meaning-making opportunities. Anita preferred to access Yahoogroups via the Web, for instance, rather than use her University of South Florida e-mail account: "I don't take chances with getting all those ads for sex and porno and perverts even if it's a class thing. Picking the Web Site way was the only way I want." Were the technology utilized listserv, listproc, or majordomo, Anita would not be able to make this choice; Yahoogroups, however, makes it possible, as do Lyris and other like systems. At the same time Yahoogroups makes Web access possible, it also appends an advertisement to every message sent to the list and directed to subscribers' e-mail accounts, clearly a problematic push initiative. Jose agreed with many in the class who felt the advertisements were not appropriate: "Should I be learning stuff or buying stuff? It's like reading the Sunday paper and having to go through all those ads. I don't want to read that." In many respects, I agreed with Jose and other class members who resented this sort of commercialization in a push format. I chose Yahoogroups anyway because of the Web interface and other features it offers, like a calendar and file uploading, but I do not in any way believe it is an ideal technology and/or system. We did discuss in class how to avoid some of the problematic implications of the advertisements, however, and this activity seemed to help. We talked about accessing e-mail via a telnet or other basic e-mail system, where the advertisements were attachments, rather than full-color displays, for instance, and students did seem to appreciate the critical engagement of the problematic issue, though often still disagreeing with my choice of technologies.

Considering virtual systems requires specific attention to access, an issue spanning both real and virtual influences on the systems. When many scholars think about access, they imagine the chief challenge as to provide computers and other technologies in real systems, but the reality is that any access issues in real systems are multiplied in virtual systems. Josh experienced access issues in trying to access messages from the class e-list, for instance:

> I have a cheap ass computer. It sucks. It just sucks. I got online ok, but it was just so dam slow. When I started getting messages at the site I thought good I can get things done now. I wanted to write back about . . . [Susan's] message and I wrote my email and everything but then how could I send it? There wasn't a button or anything to click on at all. I was really pissed. My message was good, but nobody got to see it.

Josh experienced two common and significant virtual access issues. When he explains that browsing the Web is slow, he's talking about the speed at which individual pages load, and the slowdown could be caused by any of several factors or a combination of them: slow modem creating lags in data transmission, slow processing speed in reading data into a Web browser, or Internet congestion. The other access issue Josh experienced relates to his inability to find a button that will enable him to send the message he has composed to our class e-list via the Web interface. Put simply, Josh's browser must have been dated, unable to display and operate forms like the Yahoogroups site utilizes. John was not the only student to encounter access issues relating to our class e-list; in fact, I heard from seven students, each asking if I could make some changes or otherwise do something proactive to help. Responding to the students' concerns, I budgeted more class time for students to use, knowing that the classroom's computers would enable all students to access and send message to our e-list, and I also assembled and distributed a master list of all open-use computer labs on campus, including their hours of operation and phone numbers. In this way, our collaborative addressing of virtual access issues enabled us to make the e-list a more equitable space in terms of individuals' opportunities to participate actively and consistently over time.

Examining the class e-list as a virtual system enables us to learn important information about meaning-making. We learn first that virtual identities prove complex, especially in scenarios where individuals know each other in both real and virtual spaces. Virtual identities in such contexts incorporate both real and virtual influences. We also learn that technologies matter, even in virtual environments where it would be easy simply to accept whatever technologies create a virtual environment, rather than analyzing the choices. Last, we learn that access issues manifest themselves strongly in virtual spaces, shaping and even sometimes controlling the opportunities individuals have to make meaning.

Cyborg Literacy: Integrated Systems

While the individual systems mentioned prove interesting, integrating them into a construction of cyborg literacy makes them all the more important. Like the choice of systems to study, the means by which integration occurs is dependent on the judgment of individuals studying meaning-making. We might integrate systems into an understanding of a single student's or single group's meaning-making, for instance, or into an understanding of how individual technologies shape meaning-making. We could also explore the influence of other system elements, like furniture or climate control. When an individual integrates meaning-making systems, then, into a construction of cyborg literacy, the integration must have a specific intent in order to be useful. At the same time I am suggesting the integration stage requires unique choices, however, I also want to suggest that it is intimately linked to earlier choices made, thus that all choices are connected in consequential ways. If I know that I ultimately want to understand more about how the set-up of a classroom influences meaning-making, for instance, I would choose to study systems that promise to offer information in that regard. Were I instead interested in learning about a particular student's meaning-making, I would choose different systems to study.

In the particular case study shared in this chapter, my goal was to identify and catalog all individuals, technologies, and other elements in interesting meaning-making systems, rather than emphasize one or even a specific instance of one, so in many respects, this case study has been nonstandard. Indeed this sort of approach might not ordinarily yield much of significant value about meaning-making because it is so general, lacking the sort of specific focus scholars would normally use to select systems and then integrate them into an understanding of cyborg literacy. In the context of this chapter, however, the general identification and cataloging is telling because my key claim about cyborg literacy is that it is broader and more inclusive than other terms developed. The identification and cataloging enables me to illustrate that breadth and inclusiveness, then, as the following table integrating elements of the various systems shows:

	Real System One: The Workstation	Real System Two: The Classroom	Virtual System One: The Class E-list
Individuals	Jose	20 students, 1 teacher	21 virtual persons
Technologies	Dell computer, cellular phone, portable CD player, headphones, and AOL Instant Messenger	35 desktop computers (27 Windows 2000, 8 Linux), 3 printers, 3 television sets, a VCR, a DVD player, 3 servers, 2 telephones, and a flatbed scanner	Yahoogroups, Unix mailer, and Web browser
Other Elements	Hard plastic chair, small table, three notebooks, diet soft drink, near-empty book bag, and sandals	35 workstation tables and chairs, bookshelf with dictionaries and thesauri, 2 white-boards, a chalkboard, and a small coffee table	Access issues, like availability of hardware and software, speed of connection, and familiarity with Internet-based media

In sum, my examination of cyborg literacy cataloged 42 distinct real and virtual individuals, 56 technologies, and 41 other elements

What matters most about the table as it reflects my intentions for the case study is the number of interesting individuals, technologies, and other elements it possesses because they are the building blocks with which a scholar would construct cyborg literacy relative to a more specific focus. Someone who wanted to learn about Jose's meaning-making and chose the same systems I did as important would be able to incorporate information not just about Jose and his specific real and virtual meaning-making practices, for instance, but also his relationship to others in real and virtual systems, the others' own meaning-making practices in the same spaces, the way specific technologies influence meaning-making opportunities in those spaces, and what other real and virtual system elements like furniture and climate control could contribute. Such breadth and inclusiveness empowers scholars to learn more and more about meaning-making by integrating systems into constructions of cyborg literacy.

CONCLUSION

As this chapter has sought to demonstrate, cyborg literacy offers the computers and writing community a new look at contemporary meaning-making. Its strengths are simple: the way it makes visible how meaning-making systems including individuals, technologies, and other important elements in shared contexts may be integrated into a detailed look at meaning-making and the way it empowers us to make decisions freely about what systems are most important at the same time it requires us to be careful and responsible. What we learn and know about meaning-making in this cyborg era is up to us, finally, if we adopt cyborg literacy as our concept because of the diversity and inclusiveness it encourages and enables.

Community Voices

Ted Nellen

Why do you choose to be active in the computers and writing community?

I have chosen the C&W community because the computer medium has provided the best vehicle to promote scholarship for all. Scholarship is the practice of inquiry into our practice as teachers and students. The three tenets of scholarship are to make your work public, to engage in peer review, and to pass it on. Heretofore this was just for the PhD candidate. Why? I found with computers I could practice scholarship for all young scholars as they could publish their work on the WWW and then engage in peer review. This is the best of the writing community and the computer makes it happen.

Since I began using computers in my NYC high school English class in 1984, I have found so many benefits for the young scholars from equity in printing out all papers, to sharing them electronically with peers in class via the network, to publishing on the WWW and sharing with the world. Computers are more than a glorified typewriter, but are magnificent publishing tools. I am reminded of the statement: "Power to the press and he who owns one." All scholars now own a press when they own a computer with Internet access. Computers and writing are a natural combination and I have enjoyed pursuing taking advantage of this power for the sake of my scholars.

Robert Nideffer

What worries you about the computers and writing community, and why does it worry you?

That it struggles to differentiate itself from the plain old "writing community." It's like asking "What's the most important aspect of the 'ink quill and writing community,' or the 'ballpoint pen and writing community,' or the 'typewriter and writing community.'" I've yet to be convinced of any fundamentally different writing methods or practices emerging since the advent of personal computing, word processing, and the hyperlink. This struggle serves only to reinforce the same intra- and inter-disciplinary divisions that burden so much of academia, preventing not only an acknowledgement and awareness of historical context, but a more productive exchange of ideas as well.

Catherine Null

What's the best lesson you've learned from the computers and writing community, and why is it the best?

When I began a Master's in Composition in the mid-'80s, the paradigm was "discourse communities." I remember struggling with North's just completed book on the field of composition—the division of composition scholars into "discourse communities." As I struggled to understand the divisions into discourses, I realized that community was the more important word. What I have learned from being involved with the computers and writing community is that discourse about writing and community has been way ahead of the curve in the field of composition compared to many other disciplines because our field is writing—writing and writing about writing. Composition people embraced word processing, the Internet, the World Wide Web long before many other disciplines. I believe that is because composition people know the value of developing community as a vital and integral part of the learning process.

Gian S. Pagnucci

How did you come to be active in the computers and writing community?

Forget the research questions, James, and let me tell you a story. It was 1991 and I had never heard of the Internet or e-mail. My only real connection to technology in those days was the Apple II+ computer I owned, and I only used that to play Breakout and Lode Runner. I was tutoring this high school kid. I'm going to call him Dustin. Dustin was being home schooled, but he worked once a week with me on his writing skills. And he hated it. To him, I was school. We did a lot of freewriting together and he would work on stories which I would read, and we would sort of talk during the sessions, though it was mostly me who did the talking. I even gave him homework, though he rarely did any of it. It was not going well.

Then one day I saw that he was reading this book with a blue and red digitized face on the cover. The book was William Gibson's *Neuromancer.* We talked about the book for awhile and before long I was reading a copy:

> He closed his eyes.
> Found the rigid face of the power stud;
> Please, he prayed, now—
> A gray disk, the color of Chiba sky.
> Now—
> Disk beginning to rotate faster, becoming a sphere of paler gray.
> Expanding;
> And somewhere he was laughing, in a white-painted loft, distant fingers
> caressing the deck, tears of release streaking his face. (p. 52)

I fell into that book and when I came out, the world was a whole different place. Dustin and I would read the book each week. We eventually learned that this type of technology-heavy science fiction was called cyberpunk. We spent hours talking about the book and it's computer geek hero, Case. We wondered what cyberspace would look like, what it would feel like, how you would move around in it, and, of course most importantly, how you would have a fight with the bad guys in cyberspace. Pretty soon we started writing our own cyberpunk story together. And that was the key to it all: connecting, linking up, or, as Case would say, jacking in.

What the Computers and Writing community is really all about is collaboration. For my own part, it was meeting sometime later my good friend Nick Mauriello, learning together about web pages and how to teach with computers. Then building online communities for sharing student writing. One kind of technology-based collaboration after another. We co-wrote a lot about all of this and presented about it at CCCC and Computers and Writing and even did two articles for *Computers and Composition.* And we met lots of new colleagues and helped lots of students to share their papers with people all over the world.

Collaboration. That's what we were learning, the power of collaboration. That was the key. Somewhere, somehow, somewhen, we plugged in, and the world spun faster, and we all began to weep and laugh, together.

Mike Palmquist

How did you come to be active in the computers and writing community?

I was introduced to the computers and writing community during the early 1980s while I was a sales clerk working at a running store near the University of Minnesota's Minneapolis campus. A tall man with large feet came into my store on a regular basis, not because he was a runner, but because he knew we had New Balance shoes that fit his large, wide feet. He was a professor at the University and seemed to enjoy a good conversation. It wasn't long before he was a regular, stopping in even when he wasn't in the market for a new pair of shoes.

In 1989, after I'd been in graduate school for three years, I filled in for my advisor, Chris Neuwirth, at a meeting in Washington, DC, where we were discussing progress on Trent Batson's ENFI project. During a break, I found myself standing in the men's room next to a tall man with large feet who seemed familiar. I looked over at him and asked, "14 Quad E's?" Terry Collins laughed and said, "GBS Sports?" I asked him how he'd been doing with his shoes, and how he'd come to know Trent, and he asked me how I'd become part of the computers and writing community. Looking back, I realize that I had found my disciplinary home before I'd bought my first computer in 1983 and long before I'd decided to attend graduate school.

Veronica Pantoja

What's the most important aspect of the computers and writing community for you, and why is it so important?

What I have enjoyed the most about the computers and writing community is just that: the sense of community. I know the term is often overused, but getting to know the people involved through conferences, listserv discussions, and other projects has been easy and pleasant. It's obvious that the people in the community work diligently to keep in touch with each other in ways that extend beyond just their scholarship. They certainly enjoy sharing and discussing their research, but, perhaps more importantly, they also enjoy each other and in inviting others in as well.

Elizabeth R. Pass

What's the best lesson you've learned from the computers and writing community, and why is it the best?

I was in graduate school in the early 1990s, and I've now had the benefit of teaching and researching in a number of (overlapping) areas: speech communication, rhetoric, composition, computers and writing, technical and scientific communication, and Web design. But what I find particularly outstanding about the computers and writing field is the willingness of its members to listen not only to its mentors and peers but also to those just entering the field, graduate and undergraduate students, and those outside the field.

We've all learned it's a challenge to keep up with the cutting-edge technology; as I grapple with the technology, I depend often on my students as teachers. Together we teach each other and as a whole we just might keep up with much of what's out there. As a graduate student in the computers and writing community, I watched as mentors welcomed their graduate students' opinions. I was excited as peers from other universities created collaborative projects and interclasses.

The lesson I learned from the computers and writing community was to value collaborative learning. I learned that to make collaboration work, the teacher or community needs to create the environment for those in different power roles to feel free to share. The computers and writing community had certainly created the environment where every kind of person could add to the body of knowledge; the question was: could I do the same in my classrooms? I use the computers and writing community as my model for the environment I want to create in my classrooms each semester. I want to learn from my students in the Online Publications Specialization courses and other courses and hope that I am creating an environment where they learn from one another and me. I am just as much a student, as I am a peer, as I am a teacher to the other students/peers/teachers in my classes. And as a teacher, that is the best lesson I could've hoped to learn.

Nancy Patterson

How did you come to be active in the computers and writing community?

My entrance into the computers and writing community began as a requirement for a rhetoric course I took for my graduate degree program at Michigan State University. The professor of the course, Dean Rehberger, asked students to subscribe to a rhetoric-related e-mail discussion list. For a number of reasons my knowledge of critical and cultural theory was quite limited and I knew I needed to catch up. So instead of joining just one e-mail discussion list, I joined several, hoping I could lurk and learn. The most vibrant of the lists I joined was the old ACW-L which has now turned into TechRhet hosted by Interversity, and deals with issues related to computers and writing. I found here a nurturing community that seemed to thrive on taking risks in the classroom and that warmly welcomed newcomers.

At the same time that I was lurking in this computers and writing discourse community, I was also reading theory about electronic text—George Landow, Michael Joyce, Jay Bolter, among others. And with limited tech skills and a lot of hope, I introduced hypertext composing into my middle school classroom. I was astonished at not only my students' engagement, but at their abilities to discuss their textual choices. I knew that I had found an environment in which my students and I could talk about decisions we make as writers. The technology had made those processes more apparent to my students (and myself) and I was fascinated by their growing metacognitive awareness. Later, when the technology become more transparent, I marveled still at the still high student engagement levels and at the textual sophistication students seemed to be acquiring. I had not seen this kind of growth when students did most of their composing with paper and pencil, even in the workshop environment that I had established. I also marveled at how much I was learning from my students, not only about their composing processes, but about technology itself. They were fearless when it came to experimenting with various programs and experts at figuring out shortcuts and features I never dreamed existed. And, of course, they were thrilled when they could, almost daily, teach me something new. I was eager to discuss my classroom experiences with others and found that people in the computers and writing community were interested in hearing what I had to share. Thus, woven throughout my classroom experiences was the support of this community. And its support served to encourage me as I continued my graduate studies and my research. Now that I have finished my degree, the electronic conversations, as well as the face to face ones at conferences, continue to provide perspectives and support as I move from the middle school to the college classroom.

Michael A. Pemberton

How did you come to be active in the computers and writing community?

I've been playing around with computers since the summer of 1969 when I took my first computer class as an entering high school freshman. The "personal computer" I learned on was manufactured by Olivetti, I believe, and it was little more than a programmable 4-function calculator (and the size of a hefty electric typewriter). Since that time, I've learned a lot about how personal computers work and what they can do, and I've also managed to learn a little about programming. (In fact, my first "professional" publication in a trade magazine was a roulette program I wrote for the Atari 800 computer, the same computer I wrote my dissertation on.) I remained on the fringes of the computers and writing community for a long time, however, largely because at the time I was learning about and first working with PCs, no such community existed. I'd have to credit Gail Hawisher with giving me my first real introduction to the computers and writing community in 1990. Knowing my interest in computers and information processing, she was gracious enough to invite me to co-author a couple of articles with her, and she added me to the editorial board of *Computers and Composition* as well.

My longtime investment in the writing center community provided another point of entry into the realm of computers and writing, particularly with the explosion of interest in OWLs in the early- to mid-1990s. I was working to get my writing center at the University of Illinois online at the same time that Muriel Harris was getting her OWL running at Purdue, so we soon found ourselves sharing ideas and experiences and eventually decided to collaborate on an article outlining our combined sense of the developing field. The C&W community has put me in contact with a lot of people who share similar interests and from whom I have learned a great deal. Through e-mail, listserv discussions, conversations at conferences, panel presentations, and a lot of reading, I've grown to feel even more a part of the C&W community over time. It's a community with many branches, and I have found comfortable niches in more than one. I cherish my affiliation with this group, and I'm proud to call many of its members my friends.

Victoria Ann Ramirez

What's the best lesson you've learned from the computers and writing community, and why is it the best?

Some folks in education hold that computers have failed to live up to their potential as educational facilitators, as tools to enable the teacher to teach more, or better. I've discovered from conferences, my own reading, and from innovators in the field of electronic communication that it's not only possible, but ultimately most effective, to structure the writing class as a writing community. The community's headquarters is situated in the traditional classroom, but can also be found online at the course's virtual-headquarters site.

Writing and computer experts such as Joe Essid of University of Virginia, and Donna Reiss of Tidewater Community College—just to mention two of my heroes/models—have created courses based on writing and electronic communication. Their courses, and those of other innovators who understand that student-centered learning can be enhanced in an electronic medium, have served as models I have adopted in my own teaching. Each semester I expand my course "syllawebs" to better reflect and serve the class writing community. From posting samples of student writing as genre models, to publishing rough drafts of projects for out-of-class feedback, to using bulletin boards to stimulate student conversation or controversy, my course Web sites have incrementally developed into spaces where students can check class info, get help on a piece of writing, share viewpoints, learn how to locate scholarly research, and publish their essays/stories. From the examples of leaders in the field of computers and writing, I've learned that once I create the online structures, electronic communication serves as a classroom assistant that's never farther away than a click, is ever flexible, and is accessible at all times to the entire class community.

Clancy Ratliff

What worries you about the computers and writing community, and why does it worry you?

Sometimes I wish we called this discipline "Writing and Computers," and not the other way around. I worry about what I've seen as the Computers and Writing community's getting, at times, enraptured with concepts such as multimodality, hypertextuality, and heteroglossia, and not foregrounding computers and the Internet in their militaristic, capitalistic, patriarchal origins—and, for that matter, in their present and future, which corporations are commanding. For example, I like that the Web is becoming a place for women to represent and define ourselves, but I cannot help but think about women in Sudan and Ethiopia who do not have the means to make their voices heard and their images seen online. I cannot help but think, too, about the women and men in Taiwan, Korea, and China who assemble computers. I want to help make the Internet as much as possible a site for activism and social change, and I am always glad to hear about teachers who combine the technology with service learning pedagogy and Computers and Writing scholars like Cynthia Selfe who are outspoken in their criticism of technology's role in (dominant) culture.

I also worry that the terms "cyborg" and "cyborg writing" are being co-opted by those who would associate them primarily with human–computer interaction, hypertext, and collaborative writing (writing by a collective, like the Borg on Star Trek). Donna Haraway argues that cyborg writing comes from colonized, marginalized writers. It is often feminist, anti-racist, anti-colonialist, socialist, and ambiguous when it comes to sexuality. I know that the Computers and Writing community as a whole recognizes the cyborg as a postcolonialist figure, but I often wonder if marginalized people and ideologies are what really comes to one's mind when he or she hears the term "cyborg."

Donna Reiss

What's the most important aspect of the computers and writing community for you, and why is it so important?

Among many importances, the "most important" aspect of the computers and writing community for me is willingness to share knowledge and experiences about the effectiveness of computer-mediated communication with colleagues at various teaching levels and in a range of disciplines. My work in electronic communication across the curriculum is founded in the theories and practices of writing teachers, in particular, the ways they are enhancing and expanding literacy with technology. The Computers and Writing community inspires and challenges me to reconsider my own teaching of writing and writing-intensive classes. As a result, I am better able to suggest to colleagues throughout the curriculum why they should and how they can integrate writing in electronic environments into their teaching.

Ben Reynolds

How did you come to be active in the computers and writing community?

In the early '80s, I was a struggling fiction writer who retyped (as in on a type-writer) each page of my work an average of 10 times. I learned deep revision from that labor, but pretty soon I wasn't learning—I was just laboring. Then I read about working with text on a computer, borrowed a small fortune from my family for a dedicated word processor, and never looked back.

It happened I was part-timing in a program to teach college-level composition to verbally gifted junior high school students. It didn't take a rocket scientist to see that the physical labor of writing by hand was damaging the quality of their work. My first proposal for a computerized writing course was turned down flat in 1985 by administrators who thought computers were for secretaries, even after I showed them Bill Wresch's *The Computer in Composition Instruction* (NCTE, 1984). By 1993, I was full-time at JHU's Center for Talented Youth, and part of my job was running a distance education writing program, a small portion of which used e-mail. I met Dawn Rodrigues at the NCTE convention. She introduced me to the Alliance for Computers and Writing, ACE, and so on. It was a mind-expanding experience. Now days, two-thirds of our 1700 students communicate via e-mail or Web.

Jeff Rice

What worries you about the computers and writing community, and why does it worry you?

My admiration for much of the work currently done in the field of computers and writing is often problematized by serious concerns regarding the integration of technology into the teaching of writing. Much of this concern revolves around the issue of software adoption. We continue to see writing programs opt for standardized systems (often labeled "courseware in a box") like WebCT and Blackboard. Even when many departments move beyond these programs and adopt systems that allow for specific technology instruction, we see proprietary software packages like Microsoft's FrontPage offered. This is despite the economical and ideological issues at stake in such adoptions (such as FrontPage being a tool specifically geared to designing web pages for Microsoft's browser, Internet Explorer). If we are to learn anything from the history of the textbook industry (see, for example, Robert Connors' "Current-Traditional Rhetoric: Thirty Years of Writing with a Purpose," Sharon Crowley's *The Methodical Memory,* and Kathleen McCormick's "On a Topic of Your Own Choosing") it is that the publishing industry often compromises pedagogical innovation by allowing standardization and compliance to override anything out of the ordinary or too demanding on a given instructor's time. Much of this problem is less the fault of the publishing industry than the institution which overburdens its workforce with 4/4 teaching requirements, thus creating a situation where writing instructors have neither the time nor the energy to learn technological skills and programs like HTML, Flash, Photoshop, or Pagemaker. If anything, the publishing industry notices that writing instructors have little time to learn how to write HTML and require a package that will do the work for them, and in turn, for their students.

The lessons of cultural study, however, teach us the danger of such compliant attitudes. If we fail to learn how the instructional tools of our classrooms are constructed, and if we fail to teach our students the implicit problems in specific, technological constructions, like courseware in a box, then we have failed the entire culture studies project altogether. As Cynthia Selfe and Gail Hawisher warned the profession some time ago in their essay "The Rhetoric of Technology and the Electronic Writing Class": Unless we remain aware of our electronic writing classes as sites of paradox and promise, transformed by a new writing technology, and unless we plan carefully for intended outcomes, we may unwittingly use computers to maintain rigid authority structures that contribute neither to good teaching nor to good learning. My issue is not that there is anything inheritably wrong with using courseware in a box or Microsoft-type programs. My concern is that

while we may ask our students to become active critics of the culture they participate in, while we may ask our students not to take for granted their cultural, racial,
gendered, and class assumptions as "natural," while we may ask our students to
critically examine the media representations they are exposed to and the places
where they work and live in order to become active citizens and not complacent
subjects, we fail to offer any critique regarding what technological tools we use in
the classroom. Instead, we inadvertently teach our students to accept technology
as complacently as they might accept any dominant discourse prevalent in the
culture (including racism or sexism). The interface of the technology remains
unquestioned and posed as "natural" (I feel compelled to repeat this term I borrow
freely from Stuart Hall). In this sense, Cynthia and Richard Selfe's description of
the politics of the interface rings true, for such interfaces are "partly constructed
along ideological axes that represent dominant tendencies in our culture." The
long term dangers of such thinking vary from a content-less computer classroom
where student empowerment disappears in favor of allowing the program to do all
the work to a vocational English department where students are only taught how
to use the software and not what it rhetorically can be used for. In order to avoid
such situations in the near future, I hope the computers and writing community
will take stronger positions regarding the movement to standardize computer-
aided writing instruction. At the very least, and in situations where instructors
have no chance but to follow the university or college mandated adoption policy, I
hope instructors will problematize the software in their classrooms through active
discussion and questioning with their students. If not, we are guilty of teaching
our students nothing more than how to type, format, open, and save.

Rich Rice

What's the best lesson you've learned from the computers and writing community, and why is it the best?

Teaching is fun, isn't it? It's exciting, challenging, and rewarding, but when it comes down to it, it's just plain fun. Eric Crump taught me this. Bill Condon taught me that effective teaching is interactive teaching. And Cynthia Selfe and Joel English taught me the importance of functional, technological literacy. I've learned so much from Web Newbold, Michael Day, Becky Rickly, Susan Lang, Fred Kemp, Judi Kirkpatrick, Dene Grigar, and so many others, as well. The best lesson I've learned from the computers and writing community is that we are *all* on the same team, and we have fun doing what we do. Teaching and learning, and helping others teach and learn, is fun. The community is an extremely generous one, and provides support swiftly, with multiple perspectives and varied experiences. You know, the computers and writing community is like one large, modem pool. Except you can dial-in at any hour, the access is always quick, and you rarely get a busy signal.

Rebecca Rickly

What worries you about the computers and writing community, and why does it worry you?

The computers and writing community has been a home, a refuge, a place without walls or institutional confines that welcomed me and gave me a sense of value and belonging. I felt a kinship with others who wanted to learn how technology, rhetoric, and teaching could be intersected so that all of us (and our students) were better for it. I also loved stepping back and looking critically with my peers at our products, examining what we'd done, what we'd found, and how we presented it. Yet I fear that these strengths—the strong community, the like-mindedness, the sense of inherent value in what we do and produce—might also be the downfall of computers and writing. Let me explain.

By necessity, computers and writing began on the margins, a community populated by those less enfranchised. When academics with little cultural capital in the academy bond together on the margins, those ties are strong; but now that the term "computers and writing" almost seems archaic (who doesn't write with computers nowadays?), I fear that these close ties become exclusive, less permeable, and they *need* to be permeable now, more than ever. If we are to remain a community, one that has moved from the margins to the mainstream, we must be more open to new ideas, new applications, and new re-examinations from those outside of our community. We need to look to other fields: technical communications, rhetoric, computer science, sociology, psychology, education, business, and so forth, to see how our field might grow and be integrated into others. We are uniquely situated to bring value to other disciplines if we enter into a dialogic exchange with them; if we maintain closed boundaries, we cannot do so, and we remain closed, ironically placing ourselves back on the margins.

A second fear I have is related to the first, and it, too, is based on our greatest strength: community. If my work with technology and rhetoric wasn't valued institutionally, I knew that the computers and writing community would find value in the work I did, by giving me encouragement, feedback, and even publication and presentation venues. I worry, though, that junior professors might take solace here, enjoying the value and even prestige that comes from this community, but NOT situating what they do in terms that their specific departments and institutions value. The computers and writing community is taking steps to make sure this doesn't happen (see, for instance, the P&T 2000 issue of *Computers and Composition,* guidelines for P&T approved by MLA and CCC, as well as the P&T Web site constructed by Cindy Selfe and a host of others as part of 7Cs), but we need to make sure that we listen to this advice and apply it to our specific institutional situations, rather than taking refuge in this wonderful community only.

Dawn Rodrigues

How did you come to be active in the computers and writing community?

I realized, early in the 1980s, that word processing was going to have an amazing influence on the learning and teaching of writing. The possibility of working with writers in a workshop setting intrigued me from the outset. I enjoyed choreographing possibilities for teaching in computer classroom settings, and because I was at Colorado State University, where we had just received a new computer classroom, I had an opportunity to be involved in pedagogical experiments with my colleagues. Working on my early NCTE publication *Teaching Writing with a Word Processor* (Dawn and Raymond Rodrigues, NCTE, 1986) was an exciting experience. I was able to develop ways of teaching students to use the word processor to brainstorm, revise, and edit their writing, and, because I was teaching in a networked classroom, I could make my lesson files available to teaching assistants who could adapt my activities and modify them to fit their own classrooms at my university. The NCTE publication gave me a chance to share my teaching ideas with others in high schools and classrooms across the country.

Although the Web offers so much more in terms of composing and publishing, my early work on word processing in a networked classroom helped me see how each teacher can contribute to the field by working with his or her students and by sharing that work with others. I continue to enjoy that atmosphere of sharing that the computers and writing community has developed and sustained over the years. Re-thinking the teaching of writing through the lens of new technology keeps me active in the computers and writing community.

Today, classroom time in a lab setting is still valuable, but more for providing students with opportunities to work on collaborative Web publishing projects, not just on composing documents. Still, the insights that teachers gained during the early years of teaching with word processors continue to be useful to us; by watching their students compose and by helping their students learn more sophisticated ways of using the power of the word processor for revising, editing, and collaborating with others, teachers learned how to design computer classrooms.

The notion of the writing classroom as a "studio," much like an artist's studio, still appeals to me. When teachers are available as writers work, they can help students when they need help, not later, when the teachable moment has passed. How better to help writers than to be there with them at the point of need.

Raymond J. Rodrigues

What worries you about the computers and writing community, and why does it worry you?

During the early days of the computers and writing community, it sometimes seemed that everyone was developing some piece of software that they hoped would make them money. Many presentations at the computers and writing conferences appeared to be little more than self-promoting advertisements for this or that software. A few eventually did become profitable for their developers, and some of those continue to be promoted during the computers and writing conferences. Even today, many participants in the computers and writing conferences seem to be overly captivated by new bells and whistles, rather than scholarly investigations into how students learn best through the new technologies, what helps them to learn, and what implications those findings may have for the field.

The journal *Computers and Composition,* however, does include investigations by researchers into the uses of computers to assist student writing, and that keeps the field going. We can be proud when our work exemplifies the essence of the scholarship of teaching. As more universities recognize and reward the scholarship of teaching, the work of our colleagues will serve as models for the new folks entering the field. As the uses of computing in the teaching of writing continue to become more and more common, it may be that the need for a special conference and a special journal will diminish, for we have entered the mainstream of composition work. If, however, we continue to concentrate upon bells and whistles, rather than the basic capabilities that the technology offers our students, then we will regress to a marginal group fascinating to ourselves, but to few others.

Liz Rohan

Why do you choose to be active in the computers and writing community?

I choose to be part of the computers and writing community because it inspires risk taking and interdisciplinary connections. Even though I consider myself a member of the Computers and Writing community, and I use computers in my teaching, computers and writing is not my main research interest which is women's literacy, and from an historical perspective. As my Mom says, Computers and Writing is my "minor." Nevertheless, my involvement with the Computers and Writing community has had a major pay off, and helped me with this historical work in an unlikely fashion. My dissertation contextualizes the literacy practices of one woman, Janette Miller, born in the nineteenth century, whose texts are in the archives at the University of Michigan. I had been studying this data informally since I learned about it when an undergraduate over ten years ago. It wasn't until I became recently involved with discussions in the Computers and Writing community, thanks, in part, to my advisor, Gail Hawisher, that I noticed how indebted Miller was to multi-media in her private texts. This realization led me to include many photos and pictures in my dissertation and to make connections between Miller's private text-making methods and hypertext. I wrote a portion of my dissertation in Powerpoint to better organize the images I employ in this dissertation. Because of this use of multi-media and unorthodox software in my dissertation, I volunteered and was chosen to be one of the first PhD students at the University of Illinois to deposit a dissertation electronically. I owe it to the Computers and Writing community for the inspiration to take such scholarly and institutional risks with my first major piece of scholarship. I expect my ongoing involvement in this community to pay off as my career proceeds.

Temi Rose

Why do you choose to be active in the computers and writing community?

Living realities are created, maintained, and changed through use. The computers and writing community is made up of people who appreciate both the practical and the magical, the relational and the reflective nature of the transformation of the experience of writing. Using computers in every conceivable way to experiment with rhetoric, structure, and meaning, we are creating the ground from which new theories of communicative resonance will emerge.

It is a tremendous challenge to redesign the concept of communication. We have taken on that challenge, not as a group where individuals compete for glory, but as a group whose development is itself developing the medium and the practices of teaching using the medium. The organization of this book is a perfect example: We develop together, in the context of one another.

Albert Rouzie

What worries you about the computers and writing community, and why does it worry you?

My concern is that the computers and writing community, perhaps justifiably, is contributing to a general tendency in rhetoric and composition to value only (or mainly) the most cutting edge of research, thinking, technology. Our graduate students are starting to complain if the course readings are over a few years old. This is even more true in computers and writing than in rhetoric and composition generally. It struck me recently that we have created a problem for our scholarship. Like it or not, our survival in the academy remains tied to publishing primarily in paper-based journals and books. The appalling lag time of print, especially books, dooms most books to be considered "historical" by us by the time they come out. While electronic publication promises some relief from this, this will take time, perhaps a long time. The effect of this superannuation of scholarship is to prematurely erase valuable studies. Perhaps this is inevitable since our work is so tied into what the technology can enable in communication, but I think we would all benefit from questioning this tendency and asking what we are losing by giving into it. It alarms me how quickly our scholarship becomes "historical" since the cutting edge bias undervalues even what we call historical.

In terms of teaching, there are losses too. I have found that many new computer classroom instructors absolutely crumble under the onslaught of constantly new technology. The progress narrative of technology suggests that advancements in hardware and software make work easier, but many experience the opposite. If we make a fetish out of the new and cutting edge, and I think we have, we risk losing our real focus on teaching, learning, and literacy. This fetish results in an inadvertent elitism based on the economic ability of the institution to remain cutting edge. So as our community expands, I hope to see our teaching and scholarship become less tied into particular levels of techno-sophistication.

Robert D. Royar

How did you come to be active in the computers and writing community?

I became involved in the computers and writing community in the mid-1980s. A colleague (Dale Lally, who ran the University of Louisville Language Laboratory at the time) attended a conference at which he met Michael Spitzer and Helen Schwartz. Michael was setting up a bulletin board system named the "5th C" at the New York Institute of Technology. Dale got an ID for me from Michael and access to the 800 number. Through that system I met all the folks at NYIT, Trent Batson, and Diane Langston and was able to discuss my own work at U of L in setting up a VMS-based networked writing class. This was at the beginning of ENFI when the acronym still meant "English Natural Form Instruction." At U of L the classes I was setting up were connected to ARPANet and BITNET, but the 5th C was a BBS only, and the discussion on the BBS was all about LANs and how they would someday supplant the larger networks such as ARPANet, USENET, and BITNET. I believe decisions that have adversely affected computers and writing were formed during that period and influenced greatly by the debate against WANs as a communication medium. I made a presentation about this subject in Manhattan in 1992 (http://people.morehead-st.edu/fs/r.royar/non-fiction/neach.htm).

As a result of the contacts I made through the 5th C, I eventually accepted a position as an assistant professor at NYIT's Old Westbury campus. Michael Spitzer was a center director (dean) at the time, and I felt I already knew many of the folks (Gary Stephens, Kathy Williams, and John Thoms in Manhattan and Marshal Kremers, Tony DiMatteo, Michael Spitzer, and Laurie George at Old Westbury). NYIT did not get on the Internet until 1993.

As a result of later computer-based community building at NYIT, I met folks from eastern Kentucky who were using a dial in service paid for by the Breadloaf school. I learned about the Kentucky Educational Reform Act (KERA) and became interested in returning to Kentucky to work with K–12 teachers.

From my dissertation on the use of computer networks to teach basic writing, to my position at NYIT and finally with my work at Morehead State University (where I am an associate professor, Director of the Writing Center, and assistant to the Dean of the College of Humanities) the computers and writing community has played a major role. In fact, I can say that without this community I would not be where I am today.

Stuart Selber

Why do you choose to be active in the computers and writing community?

I choose to be active in the computers and writing community because it is so collegial and humane. I enjoy working with people who are hopeful and who want to make a positive social difference in an increasingly technological world. More specific, I appreciate the ways in which the community values students and their work. In fact, I have found very few communities who are as willing to see their students as teachers, in part because technology is one of the few areas where one can find an inverse relationship between age and expertise.

Cynthia L. Selfe

What's the most important aspect of the computers and writing community for you, and why is it so important?

There is culturally constructed film loop that sometimes plays in the mind of many humanist scholars when they are encouraged to think of technology, or, more specifically, of computers. Sometimes this film loop features the rampaging "monster" from Mary Shelley's *Frankenstein;* sometimes the maddening voice and the blinking lights of Hal, the computer in *2001 Space Odyssey;* sometimes the murderous cyborgs/robots/androids from such movies as *Blade Runner, Terminator,* or *RoboCop.* In each case, a central character is "technology out of control."

This film loop can prove to be paralyzing when it convinces individual teachers/scholars that human agency is impossible to exert productively or strategically in technological contexts. When teachers/scholars start to believe that such agency is impossible, they forget that that they *can* (and, really, *should*) be creative and politically active in their own uses of technology. They forget, for instance, that they *can* (and *should*) contribute a humanist vision to the design of computer-supported communication environments on their campus—especially those environments used by students use to author and/or design online communications. Or, these teacher/scholars might forget that *students* need to exert agency in electronic environments—that students often need help in learning how to shape active, productive, thoughtful, and humane, relationships with computer technology and digital literacies.

To counteract the effects of this destructive film loop, I would encourage all humanist scholars/teachers to recognize that we can no longer fully understand humans as language-using animals if we fail to study the texts that humans are making, reading, and exchanging in digital-literacy environments, Nor can we understand humans if we fail to recognize the agency and power that their literacy practices and values exert on, and within, electronic environments.

It is this important understanding, I think, that computers and composition scholars contribute to the larger profession of English composition studies. These scholars know that the study of technologies must, at its heart, involve the study of the humans who design and make and use machines. Moreover, they understand that both humans *and* computer technologies constantly shape—and are shaped by—the cultural ecology they co-inhabit.

In this context, computers and composition scholars understand that technological agency *itself* is not a matter of choice; rather, such agency is an inescapable condition of humanity. The *kind* of agency we exert, however—whether it is informed or uninformed, purposeful or purposeless, productive or destructive—may well be determined by how well humans understand their complex relationship with technology.

For me, it is this realization that makes the work we do as an intellectual community so important.

Bonita R. Selting

What worries you about the computers and writing community, and why does it worry you?

What worries me most about the computers and writing community is the ever-present danger that we fail to do enough in guiding our students *away* from valorizing their prowess with the "machine" and *toward* learning the crucial thinking and writing skills they will need in their personal and professional lives. Much of the literature in our field deals with staying alert to students' tendencies toward valorizing technology for technology sake, and, to me, winning them over to an understanding of and appreciation for how rhetorical theory works in their communication efforts (when they so often disdain theory in favor of becoming experts at technological "bells and whistles") is a victory well worth our efforts. Helping sensitize students to audience, purpose, and context is crucial to their growth as thinkers, communicators and citizens of their culture. The challenge, then, is to balance time in computerized classrooms between learning with technology and detonating the power technology has over students' engagement in their own communication processes.

Donna N. Sewell

What scholarly project in computers and writing has been most influential for you, and why has it been so influential?

Susan C. Herring's linguistic analyses of electronic discussion groups fascinated me when I first found them, particularly the studies of Megabyte University and Linguist. Since then, I look for other studies by Herring, looking forward to the way she sets up her research designs and the data she uses. I know that I'll learn more about electronic communities when I read her work, even if disagree with some of her methodological decisions. Part of the reason I enjoy Herring's work, I think, stems from having a different research orientation, being more interested in conducting ethnographic research rather than linguistic analysis. Herring brings examines gender issues, speech markers, and ethical concerns in her study, which always teaches me something new about online environments.

Victoria Sharpe

How did you come to be active in the computers and writing community?

I have only been active in the computers and writing community for approximately two years, due to my lack of knowledge about its existence. During my undergraduate and graduate studies in technical communication, none of my colleagues or professors ever mentioned that such a strong community of sharing existed. Having taught in a computer classroom and experiencing all of the joys and challenges associated it while studying for my master's degree, I only had the foresight to collaborate with other colleagues in my department about computers and writing issues.

During my first semester of doctoral studies at Texas Tech University in fall of 1999, I was introduced to the computers and writing community by Fred Kemp, Susan Lang, Locke Carter, and Rebecca Rickly. I was thrilled to learn that there were others who cared enough about computers and writing issues to form a community of sharing and was surprised that there was such a large organization in operation. It has been extremely valuable attending the Computers and Writing Conference for the past two years and forming new friendships with colleagues of the same interests. My academic outlook and classroom practices have truly benefited from the annual conference, the journal *Computers and Composition,* and from my mentors in the computers and writing community at Texas Tech University.

Mike Sharples

What's the most important aspect of the computers and writing community for you, and why is it so important?

For me, the most important aspect of the computers and writing community is that it brings many voices, perspectives, theories, and practices to a shared issue, of how to enable people to write more effectively and enjoyably with the aid of technology. When I began research into computers and writing in the mid-1970s, the relevant disciplines of psychology, computing and artificial intelligence, literary theory, and media studies were far apart. I well remember the pleasure at the first UK Writing and Computers conference of sharing ideas with authors, programmers, and media theorists, a sharing that grew into a multidisciplinary community.

There are still gulfs of understanding to be crossed, for example between the European and North American perspectives on the teaching of writing, and between cognitive science and socio-cultural theory, but we're still talking! And this synthesis is essential, if the Web (and its successor, the semantic web) is ever to become a medium for literary creativity.

5

Challenging Material Conditions and the Nature of Teaching and Learning in Computers and Writing: Cyborg Pedagogy

The computers and writing community has often promoted pedagogical innovation, employing computers and other technologies to bring significant changes to teaching and learning practices. Early community leaders like Hugh Burns, Deborah Holdstein, and William Wresch designed software to assist writers to compose as effectively as possible, whether by providing rhetorical thinking strategies or an electronic writing interface. In the late 1980s and early 1990s, pedagogical innovation centered on the development and employment of electronic networks in and across classrooms, essentially creating the infrastructure for many of the teaching innovations brought forward today. In recent years, electronic networks have remained prominent, and pedagogical innovation takes full advantage of their possibilities, including electronic mail, the World Wide Web, and MOOs, among other options. The pedagogies of the computers and writing community have become mature and have begun to emphasize the critical implications of any pedagogy, along with the possibilities it brings.

Before writing more, it's important that I introduce a definition of *innovation,* as the term will be used in this chapter. I already implied that it reflects positive advancement in any intellectual enterprise, which is generally right, but a more thorough and precise definition should be crafted. In *Diffusion of Innovations,* Everett Rogers (1995) defines an innovation as "an idea, practice, or object that is perceived as new by an individual or unit of adoption" (p. 11). His emphasis that an innovation be "perceived as new" requires it to be notably distinct from predecessors, thus compelling anyone who would imagine themselves innovative to attain a firm grounding in parallel and other related innovations before crafting their own. A particularly important aspect of Rogers's definition is that an innovation is more an idea than a specific technology, if a technology is defined in technical or mechanical terms. If we were to take a computer mouse to be an innovation, then, it's not as much the plastic and wire that matter, but instead the way a mouse enables a new and important way to operate computers. For this chapter's purposes, and even more generally, I would add an additional component to Rogers' definition: social responsibility. The computer mouse described must have been created responsibly in order to be termed an innovation; it cannot cost an excessive amount of money, for instance, because that would make it a technology only for a small wealthy population of computers users.

This chapter presents a new pedagogical innovation for the computers and writing community drawn from the challenges and opportunities before us in this cyborg era: cyborg pedagogy. I begin by defining it, tying it carefully to activism for equity and diversity and thus to critical pedagogy. I then read a specific classroom experience from a class I taught at Furman University in order to demonstrate cyborg pedagogy's promise and implications.

PEDAGOGY AND MATERIALISM: DEFINING CYBORG PEDAGOGY

Discussing the material conditions around pedagogy is not new to the computers and writing community. As early as 1985, Richard Ohmann wondered aloud about the way computers and other technologies influence curricula in his *College English* article, "Literacy, Technology, and Monopoly Capital." His particular project was to imagine the sorts of technical knowledge being taught or seemingly needing to be taught and to argue that teachers must be careful not to prepare students only for the workplace. Instead, Ohmann suggests, teachers should think about empowering students by caring about their whole-person development, not simply the way they may venture into a capitalist economy and be productive or efficient. After all, Ohmann's project argues, the cost of an education is not a sum for division by quotients of labor; instead, it is about the individual and her or his ability to think critically and responsibly as a citizen of the world. This goal may well include specific attention to computer technologies, especially as they become more and more prominent in both our professional and personal lives, but it is not ever defined exclusively by such technologies. When Ohmann wrote, computers and other technologies had not advanced into mainstream American education as far as they have today, of course, so matters are somewhat different now, especially when one considers the mission and programs of many trade and technical schools. Still, however, Ohmann's emphasis on thinking critically about the goals of curricula as they relate to computers and other technologies is an important reminder for us and a particularly meaningful early example of how scholars think about the material implications of pedagogy.

By the early 1990s, researchers in the computers and writing community had begun to examine material dimensions of their pedagogy in new and important ways. Where Ohmann had brought attention to economic and class issues, scholars like Mary Louise Gomez were expanding the scope of inquiry to include race and gender issues as well. In "The Equitable Teaching of Composition," published in 1991, Gomez argues that diversity and equity should be hallmarks of responsible pedagogy, goals for us to pursue in our classrooms and in any other teaching spaces we might occupy. When computers and other technologies are in the mix, then such goals become all the more important because the technologies may extend and amplify critical implications, rather than bring positive change. Despite claims by scholars like Jerome Bump in the same era that computers offer a unique and powerful tool for democratizing the classroom, that is, Gomez asserts, they do not offer such a panacea. In terms of scholarly discussions about the materiality of pedagogy, Gomez's work proves most important not because of its argument alone, which was echoed by more and more scholars at that time, but how it signals increasing maturity in the computers and writing community. Where projects like Bump's were largely uncritical, early 1990s scholarship by Gomez and others reflects an investment in exploring the critical implications of pedagogy.

One of the most substantive and important projects in the computers and writing community to engage materiality is Christina Haas's *Writing Technology: Studies on the Materiality of Literacy*, published in 1996. She outlines her central research question in the following way: "What is the nature of computer technologies, and what is their impact on writing?" (p. 3). Haas pursues this focus by thinking about new options in the academy for engaging the materiality of pedagogy: interdisciplinarity, which she encourages as "technology studies," and binary deconstruction, by which she means a systematic disassembling of problematic binaries—computer expert versus New Luddite, for instance—that have persisted too long in and beyond the computers and writing community. Where *Writing Technology* becomes most important is when it specifically engages materiality in empirical contexts. Haas conducts primary research studies of the way writing students plan to draft and revise projects and read texts on screen, with particular emphasis to the way the "screen" or computer display technology shapes the way such textual interaction occurs. In this research, Haas extends earlier work she completed, especially the article, "Seeing It on the Screen Isn't Really Seeing It," and work by others like Cynthia L. Selfe, who had also begun to think about the implications of the computer screen. For Haas, ultimately, meaning-making should be associated with materiality in the computers and writing community because it inherently relies on the cultures people encounter and have encountered in their lives. The way we make meaning on the screen is informed by our full meaning-making histories, not just our experiences with computers and other technologies. These technologies, that is, do not automatically offer a new space where new critical orders can be raised and challenged; instead they bring with them material conditions already facing individuals in their day-to-day lives, conditions that we must address directly and responsibly in the computers and writing community.

The cyborg era calls on the computers and writing community more than ever before to develop innovative pedagogies that emphasize such material conditions. As Cynthia L. Selfe writes in her *Technology and Literacy in the Twenty-First Century: The Importance of Paying Attention* (1999b), hegemony and oppression are large-scale aspects of contemporary teaching and learning, challenging not just what opportunities may be found in educational spaces, but also who enters those spaces and what experiences they bring back with them. Despite prominent governmental initiatives like Goals 2000 and determined effort from many of us invested in the careful and responsible use of computers and other technologies in education, access remains a critical issue. Even in situations where hands-on access to technologies is available, there may not be anyone who can work with those who would like to learn, and if there is, she or he may be carrying a prohibitive teaching and administrative load, severely limiting options. Such cases demonstrate the complications of access as well because they're only the beginning of a problematic cycle. If an individual does not have hands-on access to computers and other technologies, or has such access but not instruction or support, then that individual also does not have access to electronic discourse and conventions. Specific examples are emoticons, symbols for emotions conveyed by using computer keystrokes in specific sequences, and technology related acronyms, like ROTFL (Rolling On The Floor Laughing), BRB (Be Right Back), and PEBCAK (Problem Exists Between Chair and Keyboard). In any possible access issue, material conditions matter. It might be that the community in which individual resides believes computers and other technologies to be frivolous or perhaps more commonly that the schools individuals attend offer only limited opportunities to learn about and use such technologies, but whatever the variation, the material conditions surrounding technology use strongly inform that use.

In this cyborg era, because we understand more and more about the real and possible inequities influencing the use of computers and other technologies, we have an opportunity to design and employ innovative pedagogies that can make a difference. With this chapter, I propose *cyborg pedagogy* as an option, defining it as activist pedagogy that draws on individuals, technologies, and their shared contexts all at once to promote equity and diversity. Any number of pedagogical approaches evident in the computers and writing community have sought to make a difference, so what proves unique about cyborg pedagogy isn't its activist goals, but is instead the way it requires simultaneous emphasis on individuals, technologies, and their contexts. Beyond that emphasis, however, cyborg pedagogy may take many forms; it relies on responsible educators to create and implement variations of it that stay true to its goals and emphases. Cyborg pedagogy ultimately represents an investment in equity and diversity, rather than a prescription or recipe to be followed step by step.

Through its specific engagement of material conditions and activist approach, what I am terming cyborg pedagogy intersects with the broader concept of critical pedagogy, which has assumed a prominent position in the contemporary academy. In recent years, critical pedagogy has been developed by a broad range of scholars, but most would agree that Paolo Friere's scholarship has been most influential. Freire's (1970) landmark work, *Pedagogy of the Oppressed,* argues that pedagogy must be linked to empowerment and that it, in turn, must be linked to radical democracy. It would be unfair to suggest that the two terms are synonymous, but they are intimately linked. An empowered populace is one educated on its own terms, and individuals in such a populace likewise have the ability to enact change in their lives. Compositionists C. H. Knoblauch and Lil Brannon (1993) link these intellectual goals to classroom pedagogy: "Critical teachers develop an informed reflectiveness about the conditions both within and outside schools that impinge on their quality of life and that of their students" (p. 7). Majority teacher–scholars cannot venture into other populations under the auspices of "missionary programs" or "outreach" and develop pedagogies grounded in their own experiences and knowledge, as any pedagogical intervention they bring also has ideological capital, enforcing and perpetuating the hegemony and oppression evident in the material substance of their majority cultures. Although it would be naive to assume that a neutral model of education exists, the more serious wrongdoing is failing to examine pedagogies implemented without attention to the critical implications of their presence. Any critical pedagogy that does not lead to the empowerment of its associated learners has not, finally, been successful; such a standard is strict, but necessary.

The next section of this chapter outlines one attempt at cyborg pedagogy, an assignment that called on my Furman University students to create the "ugliest" possible Web sites, both in visual structure and in content. In reading reflectively my own assignment details and the responses of my students, I strive to identify ways that the project promoted equity and diversity by simultaneously emphasizing individuals, technologies, and their shared contexts. My role in this chapter is not simply to outline the Furman experience, but to use it as a basis for detailed exploration of cyborg pedagogy and its potential to bring to the computers and writing community an innovative activist approach. What follows is not a model or ideal approach; rather it's one attempt made amid a multitude of interesting options.

THE ASSIGNMENT: LOGIC AND PERSPECTIVES

At Furman, from 1999 to 2001, I served as Director of the Center for Collaborative Learning and Communication (CCLC), an institutional entity borne of the

writing center tradition, but also moving beyond it to include simultaneous attention to computer and other technologies and to speech communication. CCLC
first opened in October of 2000, so it remains a new innovation itself. As the floor
plan reflects, CCLC provides a range of collaborative spaces in which individuals
and technologies interact:

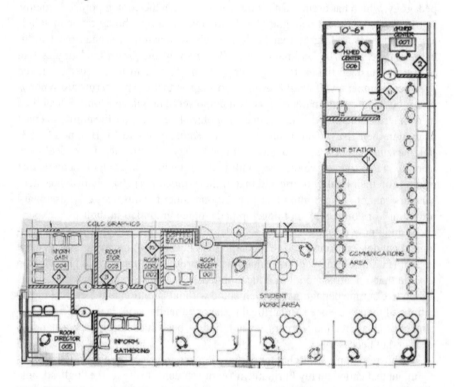

The CCLC is currently steered by computers and writing community member
Jane Love.

   One of the most significant challenges I faced immediately at Furman was how
I thought undergraduate peer consultants should be trained to work in the CCLC.
Would we have specialists, for instance, in writing, technology, or speech communication? Or instead would we have general practitioners, able to assist with
any element of the CCLC's core mission and values? Not easy questions, especially given the wide breadth of expertise necessary to support excellence in writing, technology, and speech communication together. Additionally, I needed to
decide if the training program should be a formal class experience or instead a
series of in-service training opportunities or a summer program for consultants.
I also felt a strong resistance to the idea of "training" at a liberal arts university,
where clearly the emphasis should be on the individual and the collective intellectual character of students, not on vocational preparation or other like approaches.

I ultimately proposed and won approval for an interdisciplinary course titled "Teaching Communication in Multiple Media," which the English and education departments agreed to count as an upper division course in their major and which the communication department agreed to count on an individual basis. In essence, the course divided into two equal sections: "Teaching One-to-One in Real Interactional Spaces" and "Teaching One-to-One in Virtual Interactional Spaces." Both sections emphasized planning successful one-to-one pedagogies, as well as assessing and researching them, and the virtual section included experiences with electronic mail, hypertext authoring and publishing, and MOO. Challenging students to examine and make sense of a broad range of readings, three books were required: Paula Gillespie's and Neal Lerner's (2000) *The Allyn and Bacon Guide to Peer Tutoring,* Donna Sewell's and my (2000) *Taking Flight with OWLs: Examining Electronic Writing Center Work,* and Donna Reiss's, Dickie Selfe's, and Art Young's (1998) *Electronic Communication Across the Curriculum.* Additionally, supplemental readings were assigned, including Amy Bruckman's (1993) "Gender Swapping on the Internet," Tari Lin Fanderclai's (1995) "MUDs in Education," and Anne Frances Wysocki's (1998) "Monitoring Order: Visual Desire, the Organization of Web Pages, and Teaching the Rules of Design." Over the semester, students completed four teaching simulations and a response paper corresponding to each and the World Wide Web site design project profiled in this chapter, as well as several oral reports on pedagogy and technology.

In the course, I designed a single assignment as a pilot effort toward cyborg pedagogy, hoping to learn about its promise for the computers and writing community, as well as its limitations. The assignment language follows:

Assignment Three

"Ugly" Web site Design

For this assignment you are to author and publish the ugliest Web site possible. You may define "ugly" in whatever ways you believe most appropriate, but in general, you should design your site so that anyone outside of our classroom community would believe you definitely need help with web authoring. I encourage you, in fact, to share your design with friends and to rely in part on their observations to help you enhance the site's ugliness.

Some specifics are required—in particular, your site should include at least four nodes, each appropriately developed to have value alone and in the context of the whole site. Please also make certain that your product includes at least four of the following:

• Multiple fonts, font colors, and font sizes;
• Background colors and images;
• Images in multiple formats and sizes;
• Internal and external hypertext links;
• Audio captures;
• Video captures; and
• Forms using the FormMail script introduced in class.

You may choose a theme for your site with no restrictions—it does not need to be "academic."

We will spend one class session on the presentation of these sites. You should be prepared to talk for 5–10 minutes about your work, emphasizing the ugliness of your site and the specific design choices you made in support of that goal.

Students, then, were required to construct and publish ugly Web sites, an approach I believe more challenging than calling on them to construct professional sites from various models available.

In terms of how the assignment unfolded in class, I worked to provide students with options, rather than making choices for them about how they might proceed. We began with storyboarding, and I gave students control; some prepared somewhat traditional poster boards with flowcharts and contextual notes, while others created more abstract forms. Because some of the students were new to Web site authoring, I spent a significant portion of three class periods introducing authoring options, including various software, but I did not require them to use a specific approach; rather I wanted them to have exposure to several different ways to design and build a Web site and then make choices based on what they felt would be most comfortable for them. I also introduced HTML coding, using the standard notepad in Windows and MAC operating systems. We spent 30 minutes working with each of the following applications: Macromedia's *Dreamweaver,* Netscape's *Composer,* and Microsoft's *FrontPage.* We then spent an hour developing HTML code. From there, the students selected the approach they preferred and proceeded with trying to create the Web site design they had storyboarded. I asked everyone to report their selections to me so that I could organize them into support groups, everyone working with *Dreamweaver,* for instance, and so on. These groups utilized in-class work time to confer and also e-mailed each other quite a bit, as they worked further with the assignment and needed help either with the software or with translating their design ideas into something manageable to build. I also fielded questions, but found that the students were almost completely self-directed throughout the project, even creating their own unique peer review systems.

As developed and as employed, the ugly Web site design assignment specifically represents cyborg pedagogy. I wanted to make majority students uncomfortable, to come up with a way where they might struggle as much or more than minority or disadvantaged students. The way for me to proceed, then, was to destabilize the possibly substantial access advantage majority students possessed. My base assumption was simple. If majority students had grown up with consistent web access and encouragement to take advantage of it, then they had a mature sense of what's appropriate and not appropriate for electronic spaces, and they most likely were proficient in designing high quality sites, if they had web authoring experience. At the least I assumed they would be able to list characteristics of solid web design. Asking these students to construct ugly Web sites required them to take chances, to author and publish the sorts of texts they had most likely not examined closely before. It would be idealistic, of course, to say that the assignment created equal circumstances for majority and minority students, but it was a first step, one that reveals, I believe, important information for the computers and writing community about how to proceed toward activism for equity and diversity in the future.

STUDENT PROJECTS AND SHIFTING MATERIALISM

Students ultimately responded to the assignment, as might be expected, in an array of ways. However, represented in these responses is a distinct engagement of material conditions surrounding the assignment—not always an explicit discursive engagement, but almost always a clear referencing and interrogation. In this chapter section, I use screen captures of various students' Web site designs to enable a larger conversation about the shifting material conditions evident in their projects. In particular, it's useful, I believe, to delineate majority student discomfort in completing the project, multimedia's influence on access to design structures, and minority empowerment with class conventions. Such issues will not be evident in every formulation and application of cyborg pedagogy in the computers and writing community, of course, but here they do serve to illuminate the sorts of core material issues such pedagogy must engage.

Before beginning the analysis to come, I need to offer some details about the student population at Furman University at that time because it is from this general population that my students came. According to the Fall 1999 Enrollment Survey, 2,840 undergraduate students and 613 graduate students were enrolled in October of 1999, and of these, 3,016 were White, showing a substantial majority. African Americans were the largest minority group at 192 total students, followed by Asian or Pacific Islander at 64. Women represented the majority, as their number of 2,078 equaled to approximately 60% of the total population. Additionally, the majority of students came to Furman from secondary schools in the southeastern United States, especially Georgia, South Carolina, North Carolina, and Tennessee. An interesting and relevant aspect of student life at Furman is that freshmen, sophomores, and juniors were required to live on campus, and many seniors chose to do so as well, conditions that give rise to a close-knit community, for better or worse.

In my class, I worked with 15 students. Of these 15, 13 were majority and two were minority, while 9 were women and 6 men. Students ranged in academic standing from second-term freshmen to graduating seniors, and their major fields of study ranged from English language and literature, to accounting, to music performance, together offering significant intellectual diversity for the class group. Over half of the students had even declared double majors, uniting such programs as political science and communication studies into diverse and challenging courses of study. At Furman, such creative and challenging programs are strongly encouraged, and faculty and administrators create circumstances in which the programs can be pursued. In this way, the students in my class were impressive, but not necessarily unique, in terms of their academic interests and ability to draw connections between them.

Majority Student Discomfort

Designing an assignment to be uncomfortable for majority students at an institution with little population diversity in terms of race, class, and ethnicity involved risk. After all, a critic might ask, shouldn't teachers shape assignments *to* the students in their classes, not *against* the students in their classes? Ironically, it is in this potential question that the rationale for taking such a risk exists. The phrase, "students in their classes," each time represents the majority, unveiling a logic that has oppressed minority students for years in the academy. If teachers believe they must always tailor their pedagogies to the majority, then that is much different, I would argue, than keeping the interests of those students always at heart. I did believe then and I still believe that discomfort in academic spaces, if structured carefully and responsibly, often leads to great personal and intellectual growth for students. Indeed this perspective is well supported by scholars exploring experiential learning. In "Experiential Learning Across the Curriculum," for instance, Pat Hutchings and Alan Wutzdorff (1988) feature a discussion of *dissonance,* which they describe as "throwing learners temporarily out of balance to move them towards deeper understanding" (p. 14). Ultimately, this articulation of the possibly positive influence of discomfort for learning led me to feel pedagogically and ethically secure about the Web site design assignment.

Nyah, the student whose project is profiled in this section, was a senior English major with particular interests in late 20th-century literature and feminist criticism. Her experience in my course was her first with web authoring. In completing the design assignment, Nyah seemed to battle against both years of exploring well-designed Web sites, which I anticipated, and the understanding of discourse she had formed in literature classes emphasizing a highly traditional canon, a type of resistance I hadn't expected. Nyah was unaware, for instance, of the pluralistic "english" postcolonial theorists have developed in resistance to "English," which reflects to many cultures still the imperialist conditions under which it was first introduced to them. While the Web design assignment did not require students to think about postcoloniality, it asked them to consider new and emergent forms of communication in electronic spaces, making an informed sense of discursive diversity highly useful. I'm not sure I would have realized Nyah's traditional perspective on discourse had the assignment not challenged her to try something new, an outcome that suggests cyborg pedagogy may well be able to operate successfully in tandem with other pedagogies in and beyond the computers and writing community, a coexistence that holds great promise.

In Nyah's Web site, she labeled specific sections as ugly or chose subjects that could only be reasonably construed as ugly. These design decisions demonstrate her discomfort because she rarely trusted her work to be ugly on its own terms. In the first screen capture, which is the opening page of her site, Nyah actually used the title of "Annoying":

She did not trust her design itself to be annoying, so she relied on discourse to guide site visitors. In many respects, the emphasis on discourse represents a class standard because many of the students made similar design decisions. At the same time, however, Nyah's extensive experience with language as an English major suggests that her reliance on discourse may reflect more discomfort than is evident in classmates' designs. Nyah's use of discourse might be seen as similar to an art major's use of graphics, for instance, or a music major's reliance on sound files. Discourse, then, emerges as a kind of safety net for Nyah as she completes the assignment.

Seemingly showing a different tack by incorporating an image, Nyah next introduces site visitors to "very icky, albeit very blurry vomit." After completing what must have been an unsettling web search for pictures of vomit, she included the following image:

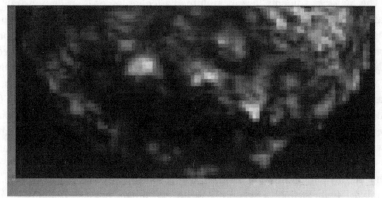

This is very icky, albeit very blurry vomit. Have a nice lunch!!

Notice, in particular, the way she has sized the image to make it blurry, a strategy that she believed would enhance the page's ugliness all the more. Despite her creative use of the image, however, Nyah felt the need to return to discourse again, featuring the text I cited just to introduce the capture, as well as a sarcastic "Have a nice lunch!" The sarcasm, I thought, did show some discursive invention, which seemed a step up from just labeling. That is, although Nyah's reliance on discourse still seems a unique reflection of her discomfort in completing the assignment, she does begin to show some creativity in her use of it, which is important. "Have a nice lunch!" is, after all, not likely to be a phrase she uses often in research papers for literature classes. Such creativity, paired with the image she selected and altered, shows both continued discomfort and an indication of progress amid it.

In student responses to any assignment grounded in cyborg pedagogy, majority discomfort may not be easy to observe. I was able to describe Nyah's particular case partly because I had the chance to interact with her in class meetings during that spring semester. Had I needed to use only the displayed screen captures as a basis for my claims about her discomfort, I would have been much less sure about my position, and I might well have not opted to take a position at all. Cyborg pedagogists in the computers and writing community, then, looking for evidence of majority student discomfort should be prepared to think broadly and holistically about the experiences and perspectives of the students involved.

Multimedia's Influence on Access to Design Structures

One particularly revealing element in student projects was the role of multimedia technologies, as majority students used them liberally while neither of the minority students did. As a researcher, I am immediately suspicious of information that fits so easily into a particular framework, the multimedia use breakdown as reflecting privilege in this case. However, the implications seem clear and powerful. The majority students' use of multimedia technologies *does* demonstrate the advantages that years of access to advanced media afford. At the same time, though, the relationship between access and multimedia authoring can be problematized. Certainly all media are not the same, and student design did not necessarily have to evolve from experience with technologies, though I would argue that some impact from privilege is unavoidable. One of the students, Brian, for instance, was a music performance major, and he developed original audio files for his site; clearly this design work was as much about his expertise in music as his experience with computer technologies. Finally what matters, then, is that, at least while the digital divide is so apparent and so telling, multimedia technologies do have strong potential for reflecting both the privilege of the majority and the hegemony and oppression imposed on the minority by that privilege.

Unlike Nyah, Arthur, a second-semester freshman, possessed significant expertise with computer technologies before beginning the class. In fact, for his fall term English composition class, he had developed two analytic hypertexts with multimedia components, these building on computer-based work he had completed in secondary school as well. Like many underclassmen at Furman, Arthur had not yet chosen a major, though he had a number of times articulated an interest in double-majoring in English and computer science. On several levels, one of the most interesting aspects of Arthur's academic life was his Army ROTC membership and training. I say several levels because I saw the influence of ROTC in his activity and performance multiple ways. First, as perhaps might be expected, Arthur was extremely disciplined; never missing class sessions, completing assignments early, leading peer groups, and more. Second, though, and much more provocative, Arthur imagined and attempted to outline a functional relationship between the computer technologies the class was employing for teaching and those that enabled the operation of warfare technologies. He talked about "plans of attack," for instance, with students in various teaching simulations.[1] Arthur was very much still finding himself, but it was clear that computers and ROTC activity strongly influenced his thinking, influence especially telling for this assignment because it brought forward his previous experiences with technologies and emphasized his access advantage.

Multimedia technologies enabled Arthur to construct ugliness in multiple ways, and he took full advantage of their capabilities, demonstrating a savvy ability to include them in site design and to think about their best uses. In the screen capture below, which is the opening page of Arthur's site, he prominently references the movie, *A Night at the Roxbury,* and does so with several media:

---

[1]I'd like to offer a fuller explanation for this claim. Arthur was in my English composition course, which was titled "Disney Culture in Multiple Media," so his hypertextual projects were written for

In addition to the image of *Roxbury*'s lead characters, he includes such text as "These guys are so cool. They get all the chicks," a link to "pick-up lines" he has uploaded, and several links to video captures of segments from the movie. Arthur also used javascript to make a MIDI audio file of the best-known *Roxbury* song, "What Is Love," played at all times, not just when activated by site visitors. Combining graphics, audio, video, and more traditional alphabetic characters, then, Arthur demonstrated significant experience with multimedia, the sort of experience that could only realistically have come from previous access to such technologies in similar contexts.

In exploring sites like that created by Arthur, it's easy to be dazzled by the use of multimedia, but because he had a great deal of design experience before beginning the assignment, the elements alone do not show that much growth. His use of them did at times, however, and I want to be sure here to indicate that majority students can show innovation in the work that they do, even if it does rely in some measure on their access advantage. Cyborg pedagogy, I mean, does not require less appreciation for majority student work; instead it provides a broad intellectual context for thinking about it in the computers and writing community and especially a means to challenge majority students by making them less comfortable in their work. The video captures on Arthur's site proved especially powerful in terms of its ugliness, for instance, as he did not have appropriate intellectual property rights to display such clips.[2] Also Arthur's participation in teaching simulations and other assignments showed he understood problematic positions like objectification, so here it seems that he is intentionally setting that tone as part of his assignment response—taking a problematic social position by constructing women as subjects of canned "pick-up lines," that is, and using it as one basis for designing the ugly Web site. Clearly Arthur has made some strides in thinking about ugliness, then, as he does not simply provide multimedia examples for his work. Instead he uses the multimedia to create problematic social and cultural situations, a tack that no doubt caused discomfort for him in authoring such texts, but that also helped him to think in new ways about the media he had used in more standard ways before.

---

me. As the last assignment in that course, students had to prepare "imagineering" proposals for new Disney attractions, and the one Arthur created was called "War." In this ride, park patrons would encounter deafening gunfire, feel threatened by violent motion simulations of battle scenes, and be forced to evacuate themselves from a sinking ride vehicle. Arthur's work, of course, was satirical in nature, but the warfare knowledge is what enabled that satire, demonstrating the influence of ROTC training and participation.

[2] I should note here that Arthur actually came to me before uploading the video captures, as he wanted to demonstrate the "ugliness" of intellectual property violations, but did not want them to be seen as his actual understanding of the issues and did not want to take any chances with public problems. The compromise we developed was that Arthur would submit his multimedia version of the project on a zip disk, not by posting it to the Web. The version posted to the Web, then, for class viewing did not include the video captures.

Arthur also demonstrates growth and development in his large-scale or meta-level design choices, these choices again simultaneously reflecting his degree of experience with multimedia and the way the assignment pushes him to think about the role of these media in different ways. Instead of relying on *Roxbury* or other popular movie texts throughout his site, Arthur chooses to interject material that has no relation at all to other parts of the site, and he does so without labeling or other context-shaping discourse. Near the end of his site, for instance, he includes a video capture of a gymnast who accidentally runs into a pommel horse, instead of vaulting up on it. A still from the video is included in this screen capture of the site:

The inclusion of gymnastics on a site auspiciously about *Roxbury* is, of course, promising for an ugly Web site design in that it interrupts site theming, but more authoring innovation enters the mix as well. Completing the assignment early, as always, Arthur had actually solicited and received peer comments from classmates, so he was able to develop a thorough assessment of audience expectations. When Arthur presented his site in class, it was extremely popular, and everyone laughed at the gymnast video, showing he had reasonably assessed their sense of humor. Arthur's example, then, simultaneously demonstrates new thinking about the role of specific media, here reflecting the idea that it can be a disruption in site theming, and a degree of discomfort with the approach, stimulating him to talk with classmates about the design choices in advance of the assignment's being due.

The way students use multimedia to complete assignments, whether grounded in cyborg pedagogy or not, demonstrates in some measure the access advantages they have had in their lives. Before beginning the web design assignment, Arthur had done a great deal of work with design in other classes and on his own, and he had included multimedia a number of times, giving him a broad base of valuable experience. It's not fair, ultimately, to assume that majority students automatically have had more access; any absolute, as postmodern scholar after postmodern scholar has shown, can be problematized. The point here is only that cyborg pedagogists in the computers and writing community need to consider access carefully in assessing their students and the way they may or may not be empowered by their work.

Minority Empowerment With Class Conventions

One of the most compelling aspects of cyborg pedagogy, and thus most interesting for the computers and writing community, is the way it empowers minorities to imagine a place for themselves in worlds rich with technology. In Freire's (1970) terms, minority empowerment involves creating educational spaces in which students can explore technological options and designing assignments that enable diverse responses, not boilerplate, to use a concept from contract law, projects that squelch all potential for alternate possibilities. Cyborg pedagogy sometimes means that teachers do not receive the sorts of responses they imagined when introducing the assignment, even if they allow for diversity, so any sort of empowerment must always be a collaboration between teacher intentions and student interests. Done on its own terms and in its own way, cyborg pedagogy can empower minority students in telling fashion.

Monica, the student whose work is profiled in this section, was a graduating senior with a double major in communication studies and political science. She was one of the most mature students in the class in that she could locate and talk about intellectual issues, like ethics and ideology, and pragmatic issues, like dress code and timeliness, at once in the same tutoring situations. Because of this maturity and a generous nature, Monica seemed to serve as an informal mentor to many of the younger students in the class. One interesting note about Monica is that she had very little experience with computer technologies before entering the class; she had used e-mail once or twice and browsed the Web several times, but nothing more. This level of access seems to parallel standard assumptions about minorities and access to computer technologies, for better or worse.

Some of Monica's design proves aligned with the class standard, I should note, instead of being particularly innovative. In the first screen capture, which is the opening screen of Monica's site, she uses both a label, "Bad Website Design," and a theme, "It's All About the Music," which she carries throughout her design scheme:

Readers may notice that the screen capture on top and those following are more difficult to read than normal, and that's because Monica used a bright green background with bright yellow font to add to her site's ugliness. The next screen capture demonstrates the use of an image:

Instead of simply including an image, Monica resized and reconfigured it such that it is blurry, just as Nyah did with her image of vomit, and Monica also again used the class standard of labeling by writing "The picture was loaded too big so the image is hard to see." At this point, some intellectual development is evident, but Monica's progress seems almost identical to that of Nyah. They both emphasized discourse, and they both included an image and resized it so that it would be blurry.

Monica later demonstrates, however, a growth and empowerment more profound than her peers. The screen capture presented now serves as one example:

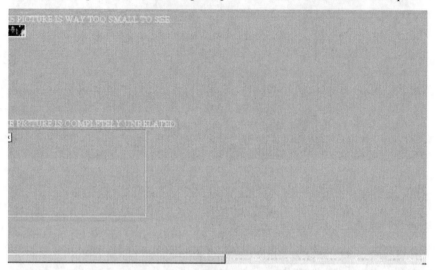

The two labels she uses in the screen capture above are the following: "The picture is way too small to see" and "The picture is completely unrelated." Again, at face value, such use of discourse seems standard. However, a closer inspection showcases Monica's empowerment. In the capture, she actually manipulates the class standard, taking the majority's reliance on discourse and altering it in a quasiparody. Under the label reading "The picture is completely unrelated," that is, she does not include an unrelated image; instead she includes a missing image error, shifting the labeling dynamic in a powerful way. The label, then, is false, a tactful misdirection that makes site visitors stop and think about their expectations. Several students in the class thought at first, for instance, that Monica had simply forgotten to upload the appropriate image file, before they realized what she had really done.

As evident in Monica's case, cyborg pedagogy opens key spaces for minority empowerment. Such a result is exciting not only because she developed a sophisticated and critical perspective on various computer technologies, but also because she also aggressively overcame the access differential that had previously informed her use of technology. Like Haraway's (1985) cyborg figure, Monica seized the technology that had oppressed her and reinvented it in a way that brought out her voice, her innovation, and her spirit. Both locally and broadly, then, Monica's response to the cyborg pedagogy reflects important growth and development.

CONCLUSION

In and beyond the student projects shared and their material capital, cyborg pedagogy must be grounded in the computers and writing community by the sense that sometimes tangible and observable changes are not the best results, if the goal of an assignment is to destabilize majority power structures, as reified through years of advanced access and extensive privilege. It follows, then, that responses to cyborg pedagogy may often not meet well with institutional parameters and guidelines, if students are given the freedom to author a broad range of responses. In this sense, such assignments are likely not the best for strict grading, departmental standards, and institutional goals and commitments. I opted, for instance, never to disclose my goals for the assignment to the students, not even after their projects had been submitted for grading. To this day, they do not know the pedagogical risk I was taking and the way I saw it fitting into their university education.

The risk-taking character of cyborg pedagogy is perhaps the most poignant challenge for would-be cyborg pedagogists in the computers and writing community. Such teachers are often in vulnerable positions as graduate students and adjunct faculty, not in tenured lines, a reality that compounds the risk instead of alleviating it. As a junior faculty member, I myself felt limited and cautious about how far I could push institutional bounds, a professional pressure that at least partially influenced my decision to formulate only one assignment in cyborg pedagogical terms. Writing this chapter, to be honest, has not eased my mind.

If teachers' first priority is the growth and development of their students and if some courage can be reasonably mustered amid high pressure environments, cyborg pedagogy offers an important and powerful option in the computers and writing community. Its connection between teacher innovation, genuine care about students and their futures, and social responsibility proves compelling in this age that offers so many challenges. This chapter, I hope, is a beginning toward more such connections in the future—not because it necessarily inspires increased risk-taking, but because it offers one means for teachers and students to share a rich dialogue about the material implications of their educational futures. Such a dialogue would be a powerful innovation indeed for the computers and writing community.

Community Voices

Michelle Sidler

How did you come to be active in the computers and writing community?

Studying under James Berlin at Purdue University in the early 1990s, I wrestled with the place of economics in composition studies. In particular, I was looking for a way to explain the materiality of writing—its systems of commodification and tools of production. Simultaneously, I began teaching HTML and web page development in Technical Writing and soon realized that computers embodied, symbolized, and in fact foregrounded the economics of writing.

I became a member of the computers and writing community because I found a group of scholars, Cynthia Selfe, Gail Hawisher, and Johndan Johnson-Eilola in particular, who not only articulated the connections between computers, writing, and economics but also argued that writing, like computers, is itself a technology with material components. For me, this field now encompasses more than just computers and writing. Those terms are short-hand for a community that analyzes the symbolic systems of communication, from writing and economics to visual, aural, and digital media to even biotechnology and the very codes of life.

Greg Siering

What worries you about the computers and writing community, and why does it worry you?

Over the past few years, the mainstream acceptance and application of computer-supported composition pedagogies has led computers and writing specialists to evolve and redefine themselves within their departments and scholarly fields. My greatest concern for the computers and writing community is that our need to "push the envelope" and stay on the "cutting edge"—in essence, justifying our specialization—may alienate us from the majority of composition instructors who may benefit from our guidance and experience. As innovators and early adopters (Everett Rogers' terms), we must balance our desire to explore new theories and pedagogies with our obligation to stay close enough to the early and late majorities to be of use. For years we have worked hard for acceptance and credibility within composition studies, and now we are enjoying the benefits of our work as we are becoming more useful to the mainstream. If we pack up shop and move too quickly to explore new and more radical areas of computers and composition, we must be ready to exist again out on the fringe again. Of more use to everyone, however, would be to work at maintaining a balance between exploring new territory and meeting our colleagues where they are today.

Geoffrey Sirc

What's the most important aspect of the computers and writing community for you, and why is it so important?

The best scholars in computers and writing are like delighted children, with a fervent, blissful confidence in their toys. The staggering energy that children have at play is the kind that seems to permeate C&W scholarship and pedagogy. I remember once, over a dozen years ago, Michael Joyce was describing his classroom to me (we were talking about our then-new use of local-area chat programs): students were all type-chatting away, the lights were dimmed, and music was blaring on a boombox. He sounded like a kid, having engineered some elaborate play-ritual for friends at his house. C&W pedagogy, then, is not thin; it's full of many layers of words, discourses, texts, lights, and images. The scholarship, too, when it's good, has that same complicated fizz to it. Plus, I have never failed to enjoy the kooky positivity of C&W scholars. Their faith in their technologies is boundless. If ordinary composition is a plutography of successful writing, C&W composition is an almost laughably absurd hyper-plutography. Any given suggested application will not only change students' writing, it will change the world—it will bring about undreamed-of increases in productivity and articulation, change the very nature of reading and writing, redefine gender roles, rectify power imbalances, and transmogrify base humanity. The docuverse is now! Who can resist such egregiously pure faith? What C&W scholars write is dream, vision. Their published scholarship is one of the last vestiges of the kind of home-made expressivism you used to find all the time in composition before it lusted after academic respectability. C&W scholars know they will never have academic respectability (like children, they can't conceive of themselves as grownups, they're both unable and unwilling). They take so much more risk with form when it comes to writing. That freshness in both the form and content of their scholarship doubtless comes from their immersion in the material stuff of composition. Those of us who work/ed with the discourse-traces technology captures stumbled upon, say, Barthes as if it was the long-lost manual for our software. Similarly, scholars today working with issues of the visual are forging theories of the digital that may or may not already have been theorized. Having to deal with the manifest realities of technologized composition seems to guarantee that raw, nutty certitude that is the hallmark of C&W scholarship, keeping it the best thing that never happened in composition studies.

John Slatin

Which scholarly project in the computers and writing community has been most important to you, and why?

I'm going to take "project" (and maybe "scholarly," too) in the broadest possible sense here, and I'm going to talk about two projects, not just one.

The first project is the effort by virtually the *entire* computers and writing community to develop an understanding of how to use computer networks, first local and now global, as teaching and learning environments that expand what "writing" is and enhance our ability to help students learn to write, to take pleasure in writing, to make writing in all its senses a powerful medium for them. That project is of course ongoing; I was most intensely involved with it from 1987, when I first taught in a networked environment using what eventually became the *Daedalus Integrated Writing Environment,* to approximately 1995, when my deteriorating eyesight made it increasingly difficult to take advantage of the tools. Even so, one of my goals as director of the Institute for Technology and Learning since 1996 has been to find ways of taking the lessons learned during my years in the Computer Writing and Research Lab here at UT, and learning how we can generate the same kinds of excitement among both teachers and learners in K–12 education.

Both of these enterprises intersect in important ways with what's now the project of most direct importance to me personally. This has to do with learning how to make the Web and other computer-supported teaching and learning environments accessible to and inclusive of students, teachers, and others who have disabilities. At the moment I'm focusing primarily on the Web, which has become so dominant in such an astonishingly short period of time. I am trying to work out the ramifications of an approach I call "AccessFirst Design," which makes meeting the needs of learners with disabilities the starting point for and first priority of design; and of what it might mean to treat accessibility guidelines and standards as design resources rather than hoops to jump through or items on a checklist. The goal is to create the richest possible environment for *all* learners and teachers.

Sarah Jane Sloane

Why do you choose to be active in the computers and writing community?

In 1986, when I was in my last year of working on an MFA (in both poetry and fiction) at University of Massachusetts in Amherst, one of my neighbors was a PhD student in computer science, from Paris, who was intensely interested in fractals and cache memory dynamics, if I remember correctly. At any rate, towards the end of the year, Dominique loaned me a 5-and-1/4-inch diskette (remember those?) of the first interactive fiction I'd ever seen: some shareware that produced a story about what it would have been like to sail on the maiden voyage of the Titanic (there's a powerful narrative telos for you). As reader, I was reduced to entering commands like "go north." I was hooked, even though the game was rudimentary and ended with a whimper, not a bang.

Upon reading this disk-based version of "Titanic," I immediately saw clear connections between the ways computers let readers play with narratives and the way the Oulipian games of Western European writers and theorists like Queneau, Calvino, and Perec arranged the writing experience and scripted the reading experience. Oulipo, the Paris-based "Ouvroir de littérature potentielle," or "Workshop of Potential Literature" produced textual experiments like Queneau's "The Story of the Two Peas," which really wasn't that much different from the Titanic interactive fiction I sharpened my milkteeth on. Oulipian stories were aleatory, chaotic, and often governed by algorithms.

Ever since, my interests in computers and writing has been primarily in the way the computer serves simultaneously as scripting tool and reading medium, as a place where story meets game and yields richer narratives for the blending.

Ilana Snyder

What scholarly project in computers and writing has been most influential for you, and why has it been so influential?

In the mid-1990s, I participated in the "Digital Rhetorics" project (Lankshear, Bigum, Durrant, Green, Honan, Morgan, Murray, Snyder, & Wild, 1997; Lankshear & Snyder with Green, 2000). Informed by the recognition of literacy as social practice, the two-year government-funded Australian study investigated the relationship between literacy and technology in teaching and learning. It comprised three principal components: an investigation of technology and literacy practices in a range of learning contexts; a study of selected key policy documents; and the development of a theoretical position.

Based on our site studies, we identified three broad patterns which we called "complexity," "fragility," and "continuity." In addition, we analysed the data in terms of four principles: "teachers first," "complementarity," "workability," and "equity." These patterns and principles were useful for making sense of the site studies; making decisions and judgments about what we saw; and formulating recommendations.

Working on this team project helped me get a handle on how contemporary economic, social, technological, administrative, organisational, and political changes are affecting the social practices of literacy, technology, and learning. It also enhanced my understanding of how these changes are altering literacy, technology, and learning and the relationships among them. Moreover, the excellent research practices modeled in the project have influenced my subsequent work.

Elizabeth A. Sommers

How did you come to be active in the computers and writing community?

In the 1980s while I was in graduate school studying English, Creative Writing, and Linguistics, I was also working for a Swedish hyperbaric physiologist, editing his numerous publications. Realizing that I was not the manuscript editor of whom he had dreamed, he decided to buy me a word processor. As soon as I started playing on the computer I had the first and only epiphany of my life, realizing this was not only changing his scientific writing but all of the writing all of us do all of the time. In fact, I realized this now-antiquated computer was probably the most important invention of my lifetime and that these machines were going to change literacy forever. Unable to interest a single professor in the English Department in my new convictions, I switched to English Education for my PhD, first studying the ways in which writers revised, then the ways in which they write, communicate, respond, learn, collaborate (as well as the ways many of us manage to avoid and thwart such activities).

Ever since I have stayed involved in the computers and writing community, still believing this is the most important, fascinating and relevant work I could do. The community provides me with a nurturing professional family, one with high standards and generous colleagues. I find the research cutting edge, the teaching engrossing, the pedagogical and theoretical possibilities endless, the software and hardware evolving faster than the speed of light, the educational evolutional cycles mercurial. And I'm still as interested as ever.

Terry Tannacito

Why do you choose to be active in the computers and writing community?

I choose to be active in the computers and writing community because it exemplifies the community I desire in my classes. Throughout my teaching career, I have believed in the power of social interaction, of community, to improve learning in general and writing in particular, and one of the results of that belief has been my use of peer response groups. So, when I began actively using computers to teach writing early in the last decade, one of the first books I read was Carolyn Handa's *Computers and Community: Teaching Composition in the Twenty-First Century* (1990). In his foreword to the book, Richard Lanham explains the essays' common theme: "The most profound changes wrought by computers in the composition classroom are social, political, and pedagogical, not technological. The success of these classes is measured by how well the students constitute their own social and scholarly community" (p. xiv). In my subsequent research on electronic peer response groups, one of my strongest findings was that the students felt they built extremely supportive communities in their groups. Further, the supportive nature of their communities led them to make many helpful suggestions to one another and, ultimately, led to a very strong association between the helpful suggestions and successful revisions. As I pursued my PhD and became more active with other teachers and scholars of computers and writing, two things became apparent of this group: they are a social and scholarly community, and they are extremely supportive of new members. I love being part of my departmental community, but it is in my computers and writing community that I feel I'm among colleagues who believe, as I do, in the power of computers and writing to bring both our students and our selves closer together: the power of community.

Jonathan Taylor

How did you come to be active in the computers and writing community?

I fell into the computers & writing community through no fault of my own. I always had a hobbyist interest in computers (I taught myself BASIC on a VIC-20 as a kid), and I always wanted to be a writer. While earning my MFA in creative writing I worked in an office supply store selling computers. Then, I somehow got signed up to teach a computer literacy class at the community college where I was working on the side.

It wasn't until the first day of my PhD at Michigan State University that I put the two together. I went to a class that was supposed to be about progressive pedagogy. The professor for that class took a job elsewhere and the course had changed to English Education & Computers, but I stayed after talking to the new prof, and then took other classes and taught WebCT seminars in my department at Ferris State University where I teach now. When I was hired tenure-track at Ferris, I was finishing a seminar in literacy at MSU and had some editorial experience with the Michigan Academy of Science, Arts, & Letters. It dawned on me that I could start an e-journal that focuses on literacy (which is the way I prefer to think about writing pedagogy) and technology. Hence, *The Journal of Literacy and Technology* was born. As a result of this experience, I also have helped in our department's new venture into offering online classes for off-campus programs.

Sue Thomas

How did you come to be active in the computers and writing community?

I became interested in computers in 1986 when I was studying for a degree in Humanities as a mature student. My undergraduate dissertation was on the relationships between people and machines, and I subsequently wrote my first novel, *Correspondence* (1992), about machine consciousness. My early work was much inspired by Marvin Minsky's work on AI and Sherry Turkle's research on social interactions with computers. In 1995 I began working on the Internet and established the trAce Online Writing Centre, based at The Nottingham Trent University. Now my personal writing and research focuses on virtuality, especially the landscapes of the Web.

Alice L. Trupe

How did you come to be active in the computers and writing community?

Though I was somewhat computer-phobic, I allowed myself to be talked into taking a seminar in computers and writing shortly after starting my PhD in 1993. Not one to do things by halves, I volunteered the same semester to teach in the local community college's new LAN classroom loaded with Daedalus even though I'd never used a computer for anything more than the card catalog. Feeling like the basic writing students in that first computer-based class I taught—and like Alice falling down the rabbit hole—I plunged into a variety of new teaching and learning situations, learning to use e-mail and a VAX Notes conference, *WordPerfect,* and Daedalus *Interchange* in the short span of a couple of months. It was so exhilarating—and so clearly it was the technology that enabled me to teach writing the way I thought it ought to be taught—that I started brainstorming for a dissertation study that would enable me to explore cyber-wonderland. That's what gave me the impetus to research and write about computers and writing.

But the help I got from computers-and-writing folks, starting with John Barber that first fall, guided my first tentative steps into a home in the community. Everyone I talked with, whether f2f, in MOOs, or on listservs, was excited about the way computer environments could explode traditional learning-and-teaching hierarchies. Instead of hoarding their knowledge to present or publish in a prestigious venue, they freely shared it with every mere beginner that came along. I never felt like an impostor, never felt I had to pretend to more expertise than I really had. So I wrote a dissertation about other nontraditional women students, who were balancing work, families, and school and trying to figure out computers, just as I was.

Gregory Ulmer

How did you come to be active in the computers and writing community?

The story of how I came to the area of computers and writing could begin as far back as my graduation from Brown University (1972) with a PhD in Comparative Literature, having written a dissertation on the Rousseau Tradition in D. H. Lawrence (directed by Bob Scholes). While researching Rousseau I encountered Jacques Derrida's *De La Grammatologie* and became part of the "theory boom" that was infiltrating into the Liberal Arts curriculum through programs in Comparative Literature. I started at Florida in University College (modeled on the Chicago Great Books curriculum), teaching the General Education Humanities sequence. Around 1974–75 I started teaching courses in critical theory in the English Department as well. This simultaneous teaching of the Western tradition and the French theorists who were deconstructing it produced an insight: the reason why the Western tradition was "closing" was that "dialectic" is not an absolute and universal truth but is relative as a core feature of the literate apparatus. Theory taught me that a technology of language and memory is part of an "apparatus" that includes institutional practices and human identity formation in an interdependent matrix. School as an institution (Plato's Academy, Aristotle's Lyceum, the library at Alexandria) and "Self" as an identity experience and behavior were as much a part of the invention of literacy as was the alphabet itself.

While this insight was sinking in, University College was abolished and I moved into the English Department full time (1980). I continued to teach theory, but my new responsibilities included general education writing (composition courses) and Film Studies. One reason for taking up film was as an outlet for my interest in the other arts acquired in the Humanities sequence. Again the simultaneous teaching of argumentative writing and media studies in the context of theory confirmed McLuhan's argument that the new media were affecting our moment the way print affected Renaissance Europe, or the way the alphabet affected Ancient Greece. The handbook I was using in my composition course was a distillation of a technological and social evolution going back to Plato's invention of the concept, and Aristotle's invention of logic. Our discipline was the heir of this tradition of invention, which even included its share of martyrs. I coined "electracy" as a name for the digital apparatus.

To further the new apparatus in my own realm I became Director of Film Studies in the English Department. During my three years in this capacity (1986–1989) I initiated an interdisciplinary degree in film and media studies. The goal was to apply media production to the study of the media (in short, to make our students "electrate" as well as literate). In the 1990s I turned my attention to digital media. The same year that my book outlining a theory of rhetoric for elec-tracy—HEURETICS—was published (1994), I started teaching in the Net-worked Writing Environment that had just opened with the help of a large grant from IBM. I now teach all my courses in one of the networked classrooms, and thus am able to incorporate into my assignments the use of our Multi-User Domain (MOOville), Hypertext Markup Language (HTML), and graphics pro-grams for designing Web sites, e-mail, Web search, and the like. This setting has enabled me to develop a rhetoric and pedagogy specifically for networked writing, in the spirit of the Japanese poet Basho's maxim: the point is not to follow in the footsteps of the masters, but to seek what they sought.

Victor J. Vitanza

How did you come to be active in the computers and writing community?

As I recall, I subscribed to the RhetNet List and started reading the exchanges and just jumped into them, wanting to know more. And eventually, as I learned more, I became a "regular." This was—still is, in my mind—one of the most interesting and gratifying discussions groups, at the beginning of my life online, that I have ever been a part of. It was like being on the road to Serendip, not knowing what was going to happen next. I loved the fact that I—an oldtimer—was immediately accepted into the discussions.

Many of the most fundamental ideas that I still have about our "coming community" developed from of my lived experiences typing and hyping with Beth Baldwin, Eric Crump, Nick Carbone, and so many others. Basically, we had a three-ring circus going. Additionally interesting people popped in like Tom Maddox and we had some very honest—in that nascent venue—differences of opinion about the space we were attempting to communicate in. I also remember discussions about the founding of *Kairos*. I think all involved with that e-publication—along with Eric Crump with his endeavors—are my heroes in this field.

I've always felt like an accidental academic, but on RhetNet I found my home by another accident—a fortunate, happy accident.

Zach Waggoner

How did you come to be active in the computers and writing community?

My activity in the computers and writing community can be attributed to one main factor: enthusiasm is contagious. As a young graduate student at Western Illinois University, I was mentored by Bruce Leland, whose passion for all things related to the field of computers and writing is unmatched. After several classes with Bruce, I too was intrigued with the dynamics of this emerging field. I began to attend the annual C&W conferences, and the real sense of community that I discovered there convinced me that I was where I belonged.

Cynthia L. Walker

What's the best lesson you've learned from the computers and writing community, and why is it the best?

The best lesson I've learned from the computers and writing community is flexibility. Teaching in a pencil and paper classroom, the biggest problem is a student who shows up to class without a writing instrument. However, he can usually borrow one. In the CW community, you may show up for class with the "perfect" assignment and have half of the computers in the lab down, along with the Internet, which, of course, is required in order to complete your assignment. Being prepared with a back up plan is crucial, not allowing these difficulties to frustrate you so much that you just cancel class, and keeping an open mind about alternative modes of presenting the same material are lessons I have learned well.

When teaching distance education courses via the Web, the same lesson applies. Students may use different word processing programs, e-mails are lost, assignments (those with extensive formatting) don't always appear as the author intends. So, again, flexibility is the key. Look beyond the poor layout and read the student's work for content. Keep deadlines firm, but understand there are exceptions. Remember that works for you in a face-to-face environment won't always work via e-mail and Web sites. Find alternative ways to achieve your goals.

Janice R. Walker

What's the most important aspect of the computers and writing community for you, and why is it so important?

The most important aspect of the computers and writing community for me is that it is a community. It some ways, actually, it's almost a family in that members (and anyone interested in computers and writing is automatically a "member") frequently argue and even fight with each other, and yet we keep coming back.

I first met the members of the computers and writing community when I was a brand new graduate student—my first semester in the MA program at the University of South Florida. I was sent to check out what this weird listserv and MOO Cafe thing were all about—someone had heard about them and knew I had the computer skills to figure it out if anyone could. Well, I figured it out, and I found people who were helpful and welcoming and who—amazingly—took what I had to say seriously! And many of these people were authors whose work we were reading in class.

When we finally met face-to-face, I felt as if I already knew these people—and they me. We hugged in elevators and hallways of the Hilton in Washington, DC, as we discovered from our name badges that we already knew each other. They were familiar with the work I had been doing on documentation formats for electronic sources and supportive of the work (even those who were working on developing alternative formats were supportive of my efforts!). Members of the community took each other at face value, based not on credentials or vitae accomplishments, but on the value of the work they were doing in computers and writing. At the SIG for Graduate Students, Mentors, and Adjuncts in Computers and Composition, it was often hard to tell who were the graduate students and who were the adjuncts or mentors! And, as the mark of true scholars, these people were excited about what they were discovering as they explored the possibilities for teaching, writing, reading, and research that the Internet made possible.

When I have a question, I always know I can post it online and someone will have an answer (or point me to other people or sources who will). When I'm working on a project, I can find people to talk about it with who will challenge me to delve deeper, and when I'm frustrated with teaching or projects that aren't going as I'd planned, I can find people online who will commiserate with me, kick me when I need kicking, offer suggestions, and generally support me whether I succeed or fail.

The TechRhet listserv offends some in our field with its sometimes petty squabbling (and sometimes not so petty!), well wishes and condolences for birthdays, weddings, graduations—or illnesses, deaths, and divorces. But like all families, these are an important part of our lives. When one of our members died in a motorcycle accident, we grieved together; when another member's wife gave birth, we were there to celebrate. And yet the list discussions also prod me to explore new areas in my field, to extend my research and writing projects, and to collaborate with others doing similar work.

Many of us are isolated on our home campuses. We may have wonderful colleagues, friends, and family, but often there are few people who understand the complexities of living and working in cyberspace. I look forward every year to the Computers and Writing Conference when I can see my friends face-to-face, but I would greatly miss the community online if it were to suddenly disappear!

John Paul Walter

What's the most important aspect of the computers and writing community for you, and why is it so important?

For me, the most important aspect of the computers and writing community is the support it offers to graduate students. As a graduate student who is both a medievalist and a technorhetorician, I can say that I do get support from both fields, but the support I receive from the computers and writing community is both more substantial and meaningful. While I know many Old English and Old Norse professors outside of my program, I am friends with many computers and writing professors. While a few medievalists have asked to read or discuss a paper that I have read at a conference, I have been asked to collaborate on projects and scholarship by technorhetoricians. While national medieval conferences encourage graduate student participation, the national computers and writing conference recognizes graduate student work with awards.

Could this all just mean that I'm a better suited to computers and writing than I am to medieval studies? It could; I still believe, however, that the computers and writing community is more supportive of its graduate students than other fields within departments of English, Rhetoric, and Communication. While my experience with computers and writing is not much different from many other graduate students within the field, it is much different from the experience my fellow graduate students have in their own fields. Their experiences are much more similar to the experience I have with medieval studies: there are professors who welcome us, encourage us, and support us. However, any real connection I have to Old English and Old Norse professors outside of my department exists because of my own professors, my mentors. This is, I know, how academia works. No professor, however, introduced me to computers and writing. Members of the community, professors within the computers and writing community, welcomed me, encouraged me, and supported me. And they have done more. Members of the computers and writing community, people who have no connection to me other than the one we have made ourselves, have become my mentors, my collaborators, and my friends. This support I receive from the computers and writing community, support I see offered to other graduate students in the field as well, a support that isn't there in other fields, is, I think, the heart of computers and writing community.

Cindy Wambeam

How did you come to be active in the computers and writing community?

In the early '90s, I taught in an English department with few computers and no specialized computer classrooms or labs. Writing teachers from the department began to recognize a need for technology in our classes, and, because of past experience with personal computers, I was soon called upon to help develop new labs. Before long, however, I was floundering. I had no idea where to begin designing and using a computer classroom; I didn't even know if there was anyone who might be able to help me. My constant concern was that pedagogy should guide our design, yet I was unsure how to combine familiar pedagogy with the changes that would come with new computer classrooms.

In the midst of my struggle, someone from the computers and writing community contacted me. He sent me one of my first ever e-mail messages, writing that he had heard I might be developing some resources, and offering his help. Within weeks, I had subscribed to e-lists and was using a MOO to interact with colleagues. In addition to my own personal mentor, I had found a community of teachers and scholars who were excited about new possibilities, who were learning to use a wide variety of technologies, and who willingly and openly exchanged information and ideas. Their excitement was infectious, and within a year I was deeply involved with the online "techno-rhetoricians" community. One year later, I attended my first Computers and Writing Conference and was able to meet most of my online friends face-to-face for the first time. Although Internet and computer technology has changed over the past decade, the excitement and innovation of this community has never disappeared. My technorhet friends and colleagues continue to be a central force behind much of my work.

Patricia Webb

How did you come to be active in the computers and writing community?

In 1994, Gail Hawisher, director of the Writing Center I was working for and studying in, used to print out e-mail messages sent to the entire body of graduate students and put a copy in my physical mailbox. She knew that I would never get the message because I refused to use an e-mail account. I, who had learned to teach writing in computer-mediated labs at Illinois State University, whose teaching and scholarly career was intimately tied to computers, resisted using e-mail because it seemed so impersonal. Today, in late 2001, I spend approximately 75% of my working time (scholarly and teaching) sitting at a computer, checking e-mail, sending and receiving e-mails from my graduate students who are taking an online course I'm teaching. What happened to change this? To change me? Why am I no longer receiving those important messages via paper messenger? Someone showed me how. Someone showed me how.

A fellow graduate student at the University of Illinois made the trek over to my house to install Eudora on my borrowed Mac and showed me how to use the program. As intuitive and friendly as Macs are (and as Eudora is), it took all but a few minutes before I was logging into my e-mail account where I found an e-mail message waiting for me. For someone who loves to see the mailman or woman coming down the street because I love to get mail, this was a whole new avenue of excitement. Now I could receive mail at all times of the day and night. And who knew what would be waiting for me when I logged in? I started getting messages. I started feeling connected. The excitement of staying connected with people through written text that was transmitted immediately and completely was the hook that got me.

I still feel that excitement when I open my e-mail account and find mail waiting for me. Starbucks tells me when their holiday blend is ready, Amazon.com offers me gift certificates, colleagues at other universities let me know if my article has been accepted, and my department chair informs me of the status of the online certificate program we're creating. Much of the good news I get on any given day is sent via e-mail. And when I can't check my e-mail for a day or so, I feel lost. I'm not sure if this transformation is a "good" one. There are critics of technology. But, I'll have to wait to read those critiques until later. Right now, I have to go check my e-mail.

Bob Whipple

Why do you choose to be active in the computers and writing community?

I choose to be a technoteacher, a cyberrhetorician—whatever I feel like calling it and whatever mood I find myself in on a given day—because I know I am in a field that is trying to make a difference now, that is taking solid scholarship and applying it, accepting its own limitations and criticisms, and growing from solid, yet exciting ideas that seem to have no end. What other field is so polydisciplinary, reaching through history, philosophy, technology, sciences, humanities, letters, psychology, sociology, et al.? Where else can so many different scholars find fertile ground for their ideas and respondents from different kinds of study?

I find the field invigorating because it consciously reminds itself of the past, critiques the present, and seeks the future; because it reaches for the loftiest grand ideas in order to affect the everyday; because it takes the glamour of technology to help the disenfranchised, disconnected, and dis-abled; and because it seems always to be asking "how?" rather than "why?"

Oh, and the people in it are pretty cool, too.

Helen Whitehead

What's the best lesson you've learned from the computers and writing community, and why is it the best?

A writer's job is often a very lonely one. I spent ten years sitting in front of a computer at home communicating with publishers by mail and phone, hardly ever seeing a living soul. Then along came the Internet and I discovered the computers and writing community online. As I communicated in real-time with writers all over the world it was not so much that my horizons expanded but that they disappeared altogether. Online there are no horizons, no geographical limits to contact. Not until then could I see how parochial were my language, my assumptions and my ideas. Metaphors don't necessarily work in a far corner of the earth.

Now I can learn from great individuals all over the world, communicate directly with those who are pushing the boundaries of writing with computers, collaborate with a Canadian, an Australian and my friend next door in the UK. It's a new way of seeing and a new way of working that enriches each of us, and I hope my writing is better for it. I am so enthusiastic about these new possibilities that I now work for the trAce Online Writing Centre (http://trace.ntu.ac.uk), facilitating more of these essential exchanges between writers.

Carl Whithaus

What's the most important aspect of the computers and writing community for you, and why is it so important?

I've been fascinated for a while with the implications of hypertext theory and MOOs for writing assessment. And so, I'd have to say that the work on hypertext, MOOs and electronic portfolios by members of the c&w community has had the greatest influence on me. Of course, talk of hypertext, MOOs and writing assessment often does not occur in the same journals or on the same panels—much of what has hooked my interest is the connections that can be made as one reads/ wanders from work on hypertext to MOOs to electronic portfolios.

C&W provides a place where these connections—collisions?—can be made and where others are willing to talk about these issues in detail. Where else does one get to talk to/about Michael Joyce (hypertext), Joel English (MOOs), and Kathleen Yancey (electronic portfolios)?

Since I'm playing with the idea of connections/collisions, I want to point out that c&w is not simply about links (deadlinks), but about interactions. This idea of researchers and teachers talking together about their work and students' work returns me to a paragraph from my dissertation:

> Hypertext theory and pedagogy have long emphasized the interactive qualities of computer-mediated compositions. These interactive qualities, and the blurring of lines between writers and readers in constructive hypertext, build on the methods of distributed evaluation. The idea of a single student working alone or in collaboration with a group but maintaining a clear sense of ownership of his or her words vanishes in dynamic, interactive hypertext environments. In *From Web to Workplace,* Kaj Gronbaek and Randall H. Trigg have demonstrated how open hypermedia systems can be implemented in professional environments, and in *The End of Books—or Books Without End?* J. Yellowlees Douglas discusses how fiction and literature are being transformed though web-based fiction and CD-ROM narratives that include interaction as central components in their textual structure. These studies as well as works more focused on pedagogy argue that reading and writing in computer-mediated environments are becoming increasingly interactive, increasingly back-and-forth propositions that require new theories, new perspective, and in the class-room setting new modes of assessment. Interactive assessment means that teamwork must be considered within an evaluation of effective communication.

What strikes me about the above paragraph in this context is that the idea of inter-action and teamwork is not only a classroom ideal but also a scholarly and research one—the c&w community presents a model of collaborative research not only for students but for teacher-researchers as well.

Debbie J. Williams

Why do you choose to be active in the computers and writing community?

I've always enjoyed seeing how members of the writing community have addressed problems that we continue to face. Virginia Woolf, for example, viewed "a room of one's own and 500 pounds" as addressing the needs of women writers; thus they would have the freedom from access to have time to write. Had Virginia Woolf been writing with technology available today, however, she might have added that this room of her own needed a lock on the door.

My own etherneted, system-connected, computerized cubicle has a locked door; yet often I am locked in my "room" with virtually everyone I know and some I don't. The very accessibility of my students and colleagues, even my spouse and children, during my work day is often a joy: I can encourage and be encouraged, teach and be taught, remind and be reminded, forgive and ask forgiveness in ways to a degree that time and proximity used to limit. However, though physical access to each other can be restricted, our virtual access threatens productivity. I often feel I've lost my ability to "invite." As Woolf's world would "guffaw, Write? What's the good of your writing?" so mine guffaws, "Privacy? Boundaries? Why shouldn't I expect to contact you when I need to and expect an immediate response?"

The value of participating in a writing community lies in our ability to share rhetorical principles with others, helping all of us in our communication decisions. Though no discipline offers a panacea for problems of access (as well as the many other issues and opportunities technology introduces), rhetoric offers us a way to engage in metadiscourse, using contextually sensitive principles, to communicate about academic issues while analyzing the means with which we have just communicated. As I enjoy the benefits of technology, I'm reminded by the writers around me of the rhetorical "locks" I may have forgotten to use.

Sean D. Williams

What scholarly project in computers and writing has been most influential for you, and why has it been so influential?

The most influential computers and writing project I've undertaken is building an online journal from initial conception through implementation. The project continues to be influential because it moves "writing" instruction out of English departments and into engineering and sciences while empowering students themselves to become creators of knowledge. One of my largest concerns about computers and writing is that we might become too insular in our interests and focus too much on our own work in the humanities. This journal, called *Synergy: The Journal of Undergraduate Research in Science and Engineering,* takes the knowledge we as computers and writing faculty have built and applies it in a new area, science and engineering.

Synergy influences me precisely because the journal is not so much about reporting science and engineering research as it is improving the quality of writing. *Synergy* accomplishes this through a unique editorial process that occurs in an electronic communication environment and involves graduate students in both writing and engineering and science. Specifically, undergraduate students submit their papers for initial review to a peer review board comprised of graduate students in engineering and sciences who have been trained first by writing faculty in Writing in the Disciplines and second by engineering and scientific researchers in the editorial processes of their disciplines. Once the paper has been provisionally accepted by this editorial board, the paper's author works with another graduate student, this time one from Clemson's Professional Communication program, to craft their language in a distance education, OWL model. The article is then published in a twice yearly edition of *Synergy.*

Synergy influences me most of all, I suppose, because students are both the gatekeepers of knowledge and the constructors of that knowledge. This notion of empowering and connecting students is a fundamental mindset in computers and writing and the support the idea has received in science and engineering so far tells me that what we believe in computers and writing is pretty powerful stuff.

Katherine V. Wills

What's the most important aspect of the computers and writing community for you, and why is it so important?

As a doctoral student, the enthusiastic mentoring I receive from my professors and cohorts is the most significant factor in my introduction to the computers and writing community. As a non-traditional student, I was intimidated by what I saw as the demands of learning to use digital technology. After all, my Master's thesis had been written on a typewriter. I cannot over emphasize the importance of positive affective camaraderie. My colleagues really enjoy sharing their techno-knowledge. For instance, I love being part of a community that e-mails the addresses of cool sites and shares the latest web authoring or streaming video upgrades. I know that I can find a colleague to commiserate about my students' web work in any departmental computer lab.

Indeed, for me, the process of writing and teaching became exponentially more collaborative with computers. I discarded Romantic notions of the solitary writer; isolationism became a choice of temperament, not occupation. Anytime of day or night, I could find another writer online, often from our own department! Enthusiastic practitioners and teachers ooze access and acceptance, not exclusion and elitism. I really appreciate being around people who love their work.

Sherri Winans

How did you come to be active in the computers and writing community?

I became active in computers and writing when my community college received a five-year Title III grant for technological improvements and I was asked to design our school's first Online Writing Center. It was 1998, and I had been teaching in computer-assisted classrooms since the early '90s but had limited experience with the Internet. As the grant's Writing Specialist, I used my release time from the classroom to study theory and to develop our site and my own faculty Web pages. I explored other online centers and educational sites, read about writing center pedagogy and online education, and studied developments in computers and writing. I quickly discovered *many* new-to-me resources for instructors of composition. Soon, I was contributing to several list-servs, attending online meetings with computers-and-writing faculty, and participating in face-to-face meetings at conferences. Now, toward the end of our Title III grant activities, I remain a regular at Netoric's Tuesday Café meetings, a participant in the yearly Computers and Writing online conferences, and someone who follows, as much as possible, the latest developments in the field. Our grant provided me with the opportunity and the time to become more actively involved. Without it, I and my local colleagues, with whom I share much of what I'm learning, would have missed out on some very valuable information, experiences, and professional growth. Throughout this process, I've often thought and talked about how we can involve more two-year colleagues in the dialogues about computers and writing. Others have expressed interest in this issue also, and I hope this professional community and its related organizations will continue to find ways to be open and accommodating to interested community college faculty.

Art Young

How did you come to be active in the computers and writing community?

My involvement with the computers and writing community began in collaboration with colleagues, Donna Reiss and Dickie Selfe. In the summer of 1995, we found ourselves at Michigan Technological University (MTU) in Houghton. I was a visiting professor from Clemson University in South Carolina, and Dickie was an MTU doctoral student in my writing-across-the-curriculum seminar. Donna, a professor from Tidewater Community College, was enrolled in Cindy Selfe's "computer camp," more formally known as MTU's workshop on "Computers in Writing-Intensive Classrooms." One afternoon the three of us were sitting at a picnic bench in front of the Walker Building discussing the connections between writing across the curriculum (WAC) and computers and composition. I mentioned how my WAC work had broadened to communication-across-the-curriculum (CAC), to include oral and visual communication. Donna and Dickie, who were already in the computers and writing community, mentioned that electronic communication was integral to WAC and CAC. As we continued talking, we came up with the moniker ECAC for electronic communication across the curriculum. Right then and there, we sketched out a book project that eventually became *Electronic Communication Across the Curriculum.*

As the three of us worked on that book project, my involvement with the computers and writing community grew. As I edited book chapters from authors all over the U.S., from Hawaii to Washington, DC, I learned how very talented teachers from across the disciplines used computer-mediated instruction to fulfill course goals related to WAC and CAC and ECAC. My own thinking about WAC theory and practice broadened and deepened, and my own teaching and scholarship was enriched by the work of my colleagues in the computers and writing community.

6

Setting an Agenda
for the Future in and Beyond
Computers and Writing:
Cyborg Responsibility

"Computers and writing" will some day no longer be an appropriate descriptor or modifier for the community we share. Both key terms in *computers* and *writing* have finite life cycles—computers being the most obvious, as surely newer technologies or classifications of technologies will become prominent. Writing, too, has a limited future; already scholars discuss *composition* and *rhetoric* as terms that have stood the test of time and thus will replace writing in due time (Welch, 1999; Wysocki & Johnson-Eilola, 1999). Beyond the terms, though, it just makes sense that change is coming. We don't have a "clay tablet and inscription" community today, after all, and that statement is not a critique of the continuing importance of early writing technologies; it's instead simply an illustration of the way things change.

In this chapter, I outline an agenda for the immediate future of the computers and writing community, one designed to direct our efforts carefully and responsibly toward positive change. The agenda emerges directly from the challenges and opportunities before us in this cyborg era, and thus emphasizes individuals, technologies, and the contexts they share simultaneously. I begin with an extended excerpt of a recent Tuesday Café about the future of the computers and writing community, inviting readers to listen to a multitude of perspectives on the future, rather than just my own or those I might cite from scholarship. Building on the Café discussion, I introduce the concept of *cyborg responsibility* to ground the agenda to follow.

A LOOK AT THE FUTURE: MANY VOICES

In February of 2001, the Netoric Project sponsored a Tuesday Café discussion on "The Future of Computers and Writing," and in the discussion, a number of important issues emerged. These included how much technical support, if any, computers and writing community members in faculty or professional writing roles should take on, as well as appropriate compensation for community professionals in various roles. Also the conversation at times reflected a critique of other community perspectives, demonstrating how members can publicly agree with each other about important issues. All of these components together make the conversation an important one for the computers and writing community because they reflect the maturity of the community. We do not all agree on key issues and have similar positions in institutions and organizations, and we have moved beyond looking at computers and ourselves as solutions, instead thinking about their critical implications, especially those economic, social, cultural, political, and historical. We continue even to debate what "computers and writing" is and how it is most interesting and influential. Looking forward, that is, we have not forgotten to look back.

For purposes of this chapter, I now reproduce an extended excerpt of the Café discussion. I do so without annotation because I believe it's important for readers to interact with both the form and content of the excerpt and because I want to emphasize that the path to the future for the computers and writing community will be shaped by many voices, rather than one or two. Readers who have not encountered MOO conversations before should know that multiple threads of conversation are ongoing at any single moment and that the nature of MOO discourse is often informal and playful, rather than serious. Additionally errors—typos and others—are common and accepted as realities of the fast-moving environment; they are not corrected or even probably noticed by veteran MOO participants. It's important ultimately to read and think about the voices for their ideas; the way the ideas are introduced may or may not be a comfortable read, but the ideas are no less important either way. With that introduction, I now present the Café discussion excerpt:

From The Netoric Project's Tuesday Cafe, 06 February 2001

Tari says, "yeah, what is up with that not knowing what you want a computer person for"
KarenL says, "here's a question again--to what extent is C&W associated with composition?"
Tari says, "how could you want something like that and not know"
Tari says, "er, not know why"
James says, "fixing keyboards and stuff, right? (;"

cin [to KarenL]: are you asking if the two are inseparable?

Kiwi [to Tari]: because people still have no idea what "we" do

Sherri [to Tari]: you could not have a clue what the possibilities are, if you're out of touch, and so want someone else to figure that out?

Kiwi nods at Sherri.

Tari says, "hm"

KarenL says, "see....I teach business & tech writing, and I'm studying advanced scientific writing, but most of the discussion I see is about composition."

Sherri smiles at Tari's hm

Kiwi [to KarenL]: me, too.

James says, "for better or worse, i try to get c&w away from rhet/comp-----i think the whole 'subfield' thing is really limiting"

KarenL says, "so, when we say C&W, do we raise the idea of composition, too?"

cin nods to KarenL, I do Multimedia & Tech Comm, but that doesn't seem to surface all that much in C&W discussions.

Kiwi thinks most people think "computers and writing" is about people who know a lot about how hardware and software. But many people don't seem to see "us" as ... well, as very knowledgeable about anything else (i.e., composition theory).

Brooke says, "There's an acknowledgment that students (grad and ugrad) want to do web-related stuff (both for comp and tech writing) but most traditional (i.e. lit-centric depts don't quite understand how this work can be scholarly."

KarenL nods to Kiwi....little do they know we're plotting to take over the world.

James thinks one way of examining this issue is to see what people see as "c&w discussions": journals, conferences, etc

Kiwi [to KarenL]: Shh! Don't give us away!

Sherri . o O (the take-over secret is out now!)

cin says, "I think that C&W has been associated with computer classrooms and using technology to teach...that's shifting to Distance Education more and more."

KarenL says, "part of the trick is that I'm located where Computers & Composition is edited."

Kiwi nods at cin.

cin says, "and many departments want a C&W specialist to train other teachers"

James nods to KarenL

Kiwi [to cin]: but distance ed is more and more being associated with course-in-a-box stuff

Kiwi doesn't like those courses that come in boxes.

James [to KarenL]: but the cool thing too is you have folks like chip bruce who are as much in ed studies as in comp, etc.

cin nods nods nods to kiwi, and as a correspondence course.

KarenL [to James]: I'm his research assistant.

Kiwi sees more connections between c&w and college of eds these days than
    with composition/English departments

Tari | cin says, "and many departments want a C&W specialist to train other
    teachers "

James [to KarenL]: cool! i didn't know

Tari says, "and you know, that's like a whole different thing"

cin [to James]: so it's more cross-curricular?

Kiwi nods at cin.

KarenL [to James]: I think it's cool, too :-)

Tari says, "you know what i think one problem is?"

Sherri sings It's a small world....

James [to cin]: if you go community, then you don't have to see c&w as compet-
    ing in that mix-----it's a group of people interested in similar things at similar
    times

Sherri says, "What, Tari?"

Brooke . o O (One of our priorities should be articulating how we differ from ed
    technology specialists)

James listens to Tari

Kiwi [to Brooke]: good point

Tari says, "CW sort of emerged, filled with people who figured out CW pretty
    much on their own"

Kiwi nods at Tari.

Sherri feels like a case in point.

Tari says, "they figured out how to run networks and use the web and write
    hypertext and html and use irc and mud and all that"

cin nods to tari

Tari says, "and then there they were, in the midst already"

Tari says, "doing this stuff for free"

Tari says, "and their departments figured it was all just free skills for the taking"

Tari says, "and"

Tari says, "and"

Tari says, "AND"

Sherri [to Tari]: interesting....

Sherri says, "and?"

Brooke says, "Something else to consider is how our use of technology often
    leads us to interaction with industry, which can make us seem too entrepre-
    neurial to be 'real scholars'"

Tari says, "the problem has been perpetuated because"

Kiwi nods at Brooke.

Tari says, "CW formed a community and invited other people in"

cin nods to brooke, which is always a dilemma in the Tech Comm field.

Tari says, "and the people who came in felt entitled to take and take, but not to
    learn anything themselves"

[8:21 pm]

Tari says, "i'm overstating, but i have more overstatements"

Sherri waits for more overstatements

cin grins

Tari says, "i'm thinking of an example from a conversation on MBU or ACW, i forget"

cin says, "so the dilemma is that we carved out a niche, created a community, but then weren't quite sure how to define ourselves to others...or how to protect ourselves?"

Tari says, "someone was talking about feeling exploited because she'd learned to write html"

Tari says, "and peopel in her department kept coming to her for help--they wanted her to put things on the web for them, or teach them every little thing, or whatever"

Sherri says, "I have to translate much of what I hear here, to fit it to waht's going on at my level, community college, but it does translate...."

Tari says, "and a bunch of alleged CW people jumped all over her claiming that she was being selfish if she didn't share that knowledge"

cin's been wondering a lot about how the popularity of C&W has waned a bit...it became really, really visible for a while.

cin says, "I don't think it was a fad, but I think others do see it that way."

Kiwi nods at cin.

Tari says, "where would anyone even THINK of saying that someone was OBLIGATED to take something she worked hard to learn on her own time and use more of her own time to give it away?"

Sherri says, "not waning yet at my cc, but with our limited resources we tend to be a bit behind...."

cin [to Tari]: I think I remember that

James nods to Tari

Sherri [to Tari]: we're with you

KarenL says, "composition courses."

Kiwi [to cin]: i think, too, that people think because they use email and have their students use word processors and WebCT in the classroom, that that's all there is to it.

cin shows slide 6 on the Tuesday Cafe projector.

* * * * * * * * * * * * * * * * * * *

Conferences & Publications

Have we fallen out of favor with the larger rhetoric/comp community? Do we still have sessions accepted at the major conferences?

Do articles in C&W get accepted by journals? Do we see fewer books/articles out there?

* * * * * * * * * * * * * * * * * * *

Tari says, "and with a groupthink thing like that going on, how could you become a legitimate profession?"

cin posts that slide since James sort of brought it up earlier...

Tari says, "how do you expect other people to value you and pay you for your skills if you don't force them to value them properly?"

KarenL [to Tari]: "Composition course, business & technical writing courses, people calling up English depts. for grammar help...."

James grins at cin, didn't peek or anything

Sherri [to KarenL]: you're saying that this is how comp operates, too? what tari's saying?

cin grins

KarenL [to Tari]: It's part of a wider trend, though I agree that C&W is an extreme

Sherri [to KarenL]: helpers, all?

cin says, "so it's all just service-oriented learning?"

Tari [to KarenL]: oh, admittedly it's an academia kind of thing, but at least when some random calls a comp department and asks for grammar help, the department members object

James thinks that's why it's important that we don't imagine ourselves always as different than, say, ed tech-----i like alliances with ed schools, etc, and i think they often give us a longer and more identifiable history

KarenL [to Sherri]: Sometimes, yes, I think so.

Sherri [to KarenL]: interesting

cin remembers that about 4 years ago, anything with computers & writing in the topic would get accepted at conferences, and publishers/journals were really hot after the topics

KarenL [to Tari]: Right, but they do it. No one says, make an appointment

KarenL [to Tari]: "and pay for it"

Tari [to KarenL]: yuck.

Kiwi doesn't think we've ever BEEN in favor with the rhetoric/comp community. Institutions have added the "technology stuff" to their mission statements. Comp departments needed to "open program, add technology"--but to many that just meant having a "computer person" around or stuffing students into a "lab" occasionally.

KarenL says, "As they would with a doctor or lawyer"

Tari says, "how much of this has to do with the whole romantic notion of the starving intellectual, suffering for art and all that?"

Brooke [to James]: I agree. My concern is the difficulty of articulating *within English depts* how our work is different than Ed, given the idiotic academic stigma against Ed depts.

Sherri says, "and this is why people can want a computer person and not know what for, right, Kiwi?"

cin wonders if things are changing a bit now that it's not just grad students being the resident "techie"

James thinks c&w work is still hot, but that the community itself doesn't hold up as core audience for a major publishing house----projects seem to need to be broader, or at least in terms of books-----into literacy studies, for example, etc

KarenL [to Tari]: "I think that it's more the idea that English *ought* to be accessible to everyone"

James nods to Brooke

cin nods to james

Kiwi says, "At my school, every FYC class meets every other week for a 2-hour lab. Many of the teachers use that time with computers to have their students either do timed writings or to print out their papers. And that's all most of them do with the computers--and they feel they've met the mission by doing that."

Tari says, "yeah, well, microsoft, i would say, believes that computers ought to be accessible to everyone, and if you call them with a question, you'd better believe you're paying"

Tari says, "they not only know their value, they know what the market will bear"

Kiwi . o O (and they're all agog at one teacher because she's using WebCT)

Sherri [to Kiwi]: ouch

KarenL [to Tari]: I agree with you

Sherri [to Tari]: so do we just stop?

cin [to Kiwi]: it's hard to move beyond the "how to" stage with computers, and start encouraging a deeper use

Kiwi nods at cin.

Tari [to Sherri]: yes

Tari says, "in a word"

cin says, "now, Fred Kemp says that the computer classroom is dead -- is it?"

Sherri [to Tari]: it's weird at a community college--always so few resources, so little help, so big a need!

cin [to Sherri]: and the sense of community, of helping others, is always at the forefront.

Sherri [to Tari]: I'm not arguing with you, though

James [to cin]: i don't think so at all------assuming death assumes advanced access for all, among other things

Tari [to Sherri]: yeah, i know, and everyone tells themselves they'll do this or that for this or that very good reason

Sherri [to cin]: yes

Tari says, "and they ARE very good reasons"

Tari says, "but"

Kiwi [to cin]: I tried teaching in a "traditional classroom"--can't do it anymore. I need computers in the classroom--all the time--even if we're not using them

Sherri [to Tari]: yes

Tari says, "it's bad in the long term"

Sherri [to Tari]: yes, yes, i know....

cin nods to james, totally classist assumption.

[8:31 pm]

Sherri says, "I'm the 'neighborhood coordinator' for WSU's Speakeasy, which
   some of us are using for our classes at WCC"

Kiwi listens to Sherri.

Sherri says, "so I'm the answer person..."

Sherri says, "and without this, we would not have access, or not much"

[snip]

Kiwi is amazed (still) when she finds out how little her colleagues actually know
   about the computers in their offices..... Even things I consider basic word processing functions....email still confuses them, even though they use it constantly.... and they have NO idea where they are on the web or what's going
   on there.... To them, what I do is ... well, unfathomable.

Sherri [to Kiwi]: yes

Kiwi [to Brooke]: exactly

Brooke [to Kiwi]: in the way that fixing your automobile transmission is unfathomable, or in the way that hypermedia is unfathomable?

Tari says, "i'm not sure about the tool metaphor"

Tari says, "my computer is a tool"

Kiwi [to Brooke]: to them, like fixing the transmission, i think

Tari says, "and this guy at work today, MAN, was HE a tool"

Tari says, "owait"

KarenL grins at Tari

Sherri . o O (ah, yes, tooools)

Tari says, "but see i don't think it's bad to consider your computer a tool"

James likes the automobile analogy Brooke introduced: will the next generation
   look at the inside of computers like we look at the inside of cars? lotsa
   mystery, little experience

Tari [to James]: wait wait

Tari [to James]: a car is totally elegant

Kiwi [to Tari]: it is a tool, but tools can transform tasks. I think many people in
   composition are too busy resisting changes in the tasks they already know.
   They don't want to learn computer stuff--because without knowing what you
   can with the tool.... you can't know what you can do with the tool.... (lack of
   vision because of lack of knowledge? someone say this better!)

Tari [to James]: it's like the best interface ever

Tari says, "you don't have to know about the guts, AND you don't have to be
   afraid that it's going to do something you can't explain"

Tari says, "just for starters"

Tari says, "and it's like all mapped in with your body"

James nods to Tari-----it's an interesting developmental moment, I mean: With every interface advance, we are further and further from the guts

Tari says, "pull the wheel right; you go right"

Kiwi [to Tari]: no, you don't have to know the guts--or how they work. Just what you can do with them. but it does help if you're not afraid to click occasionally!

James says, "i'm not necessarily sure what that means, but it's interesting, i think"

Tari summons the groundskeeper, who drops a topic sign off. She writes 'FURTHER FROM THE GUTS' on a slip of paper and puts it on the sign.

Kiwi grins at Tari.

James grins

Tari says, "it's like what the medialab people talk about with how computers are goign to disappear"

cin is suddenly having flashbacks to Foucault's _Birth of a Clinic_

Tari says, "not that they won't exist but that you won't see them as computers"

James nods to Tari

Tari says, "you won't have these clunky things that you sit in front of straining your eyeyes and beating your hands on"

cin [to Tari]: more natural, more intelligent, less intrusive

Tari says, "yeah"

KarenL says, "right...they're microwaves, phones, radios..."

Kiwi [to Tari]: i think that's already happening.

Tari says, "you know what? we have to have a cafe on what's his name's book, _When Things Start to Think_"

James [to Tari]: that would be cool!

Brooke says, "But what's wonderful about the state of technology now is that it forces us to pay more attention to text, and what text does. This is why MOOs are so fascinating to me."

Tari says, "maybe he'd come and talk to us"

cin says, "but for those of us teaching the professionals who will design these machines, write the instructions for them, teach others to use them....well, how do we approach that?"

Tari says, "or send some of his grad students"

Kiwi nods at Brooke.

James nods

Brooke [to Tari]: Wow, I haven't read that yet, but it sounds terrific.

Brooke clicks over to Amazon to find that book.

Tari says, "i'll just pop in at the medialab on my way to work tomorrow, introduce myself, and arrange it!"

Tari says, "and then the spaceship will come!"

[8:41 pm]

James grins at Tari
KarenL says, "Things That Think: http://www.media.mit.edu/ttt/"
Brooke [to KarenL]: Thanks!
Tari says, "yes, it's this whole project there"
Tari says, "and it's this awesome book"
Kiwi adds it to her reading list.
Tari says, "i am SO ranting and off topic"
Sherri files the bookmark
Tari [to cin]: you'd better show a slide
Tari settles down
James grins at Tari
KarenL says, "That's the name of the site, not necessarily the book"
Kiwi pats Tari (but enjoys the rant!)
James enjoys it too!
Tari says, "the book is _When Things Start to Think_"
 Tari says, "i just can't say the author's name to save my life"
KarenL says, "Gershenfeld?"
Tari says, "and i blew up netscape on this box 2 days ago and i've been too lazy
    to reboot"
Tari says, "GERSHENFELD YES"
Tari says, "neil even"
cin likes the idea of talking about that book (I guess I'd have to read it, then)
KarenL says, "That's his site"
Tari says, "he did that violin thing for yo-yo ma"
Tari says, "which he writes about in the book"
cin grins, and brings us back on topic with the next slide...

cin shows slide 7 on the Tuesday Cafe projector.

* * * * * * * * * * * * * * * * * *

THE BIG QUESTION:

Has Computers and Writing lost its singularity -- can it stand alone, or is computer technology just a part of all Writing/Teaching now?

* * * * * * * * * * * * * * * * * *

Tari says, "i think yes but no but that should be the goal."
Brooke says, "Here's the amazon.com link: http://www.amazon.com/
    exec/obidos/handle-buy-box=0805058745/stores/detail/
    one-click-thank-you-confirm/105-9324065-1275939http://www
    .amazon.com/exec/obidos/handle-buy-box=0805058745/stores/detail/
    one-click-thank-you-confirm/105-9324065-7275939alink:http://www

.amazon.com/exec/obidos/ASIN/0805058745/qid=981510251/sr =1-1/ref=
sc_b_1/105-9324065-7275939 http://www.amazon.com/exec/obidos/ASIN/
0805058745/qid=981510251/sr=1-1/ref=sc_b_1/105-9324065-7275939"

Tari says, "well, i guess it hasn't lost its singularity so much as it's discovered
that it's fresh out of focus and a sense of direction."

cin says, ""wow, quite a link :>"

KarenL imagines Pen & Paper and Writing; Printing Press and Writing

Kiwi says, "I think C&W is still a viable and essential area of study--but I'm not
sure others see it that way"

Brooke says, "Oops, sorry . . ."

James thinks the community still stands as unique and thinks simultaneously that
computers are becoming part of everything

cin wonders if the problem in C&W is that it has so many possible directions
to go

Tari [to cin]: i think the problem is the warm fuzziness of it all

Tari says, "wait"

Sherri [to Tari]: yes?

Tari says, "must not rant"

Sherri [to Tari]: but i asked....

Kiwi nods at cin.

Tari says, "well"

Sherri . o O (rant, rant, rant)

cin says, "is C&W an area you'd encourage grad students to specialize in? Would
you (or do you) teach specialty courses in C&W?"

Kiwi [to cin]: I think that's always been part of the problem of people's perception of C&W. It's not just one thing....

Brooke says, "One decent inroad seems to be Visual Rhetoric. This is a subject
that appears to be gaining more credibility as a rhetorical subject that can be
meaningfully pursued through web-based study."

cin says, "but there do seem to be some central texts, central personalities...
people we see all the time, and things we suggest others might read...right?"

Tari says, "there's been this whole effort to like make computers warm and fuzzy
parts of writing instruction"

James would encourage grad students to be active in c&w, would teach courses
in c&w issues, would teach a history of c&w course-------worrie s about its
becoming a narrow specialization

Tari says, "and warm and fuzzy creates this weird expectation of ease"

Sherri [to Tari]: ah

cin nods to tari, and that sense of ease is pretty false

Kiwi [to James]: I'd like to do that, too. I think we need people who will critically engage issues surrounding literacy and technologies

James nods to Kiwi, agrees

cin [to James]: people were surprised when I didn't declare C&W as my specialty area in my doc program (it was Technology & Society instead).

cin [to James]: but I didn't want to get too roped in.

Tari says, "and all the edges get soft and no one has any discipline about the whole business and they do things like they want their students to do something online but they don't want to work at learning it before they teach it"

James nods to cin

Tari says, "and don't even get me started on what passes for research"

Kiwi nods at Tari.

Kiwi grins at Tari.

KarenL has Writing Across the Disciplines, which tends to incorporate C&W.

James nods at Tari

Kiwi has heard Tari's research rant :)

cin says, "Literacy & Technology is one of those things that appears all over the place in job ads now -- but that hiring committees don't really seem to understand"

Kiwi [to cin]: too true

James says, "humanities computing is another growing job specialty: ohio state and maryland both had ads this year"

cin says, "it's sort of the hot phrase of the year"

Sherri . o O (humanities computing....)

cin says, "as someone who advises grad students and plans classes, I'm really interested in what sort of advice to give."

KarenL says, "Isn't humanities computing often figured as tech support?"

Brooke says, "Everyone wants a "Valley of the Shadow" type site, and they figure a humanities scholar can do that."

cin says, "I'm not sure that I should tell people to specialize in C&W -- but I certainly encourage them to publish in Kairos, present at CWC, participate in online communities."

James thinks it's associated with HUMANIST: Willard McCarty, the Virginia humanities computing initiatives, that sort of thing

[8:51 pm]

cin [to KarenL]: that's how I would have taken it

KarenL says, "Stanford just advertised for humanities computing/tech support combo."

cin says, "so is the problem just one of naming?"

cin shows slide 5 on the Tuesday Cafe projector.

* * * * * * * * * * * * * * * * * *

Is a name change in order? Techno-rhetoric? Educational Technology? Multimedia Studies?

* * * * * * * * * * * * * * * * * *

James imagines humanities computing means hugely different things for different committees/departments

KarenL nods to James

James says, "the ohio state and maryland ads were tenure lines----dunno any more, though"

Brooke says, "the cool thing about 'humanities computing' is that is sounds like we are still embracing literature"

cin nods to brooke, a professor here (who works a lot with computers & literature, and writing about literature) made a comment that sort of startled me a few weeks ago...

James thinks the name's cool as long as it's a community identifier------worries more and more about it as a field, especially if it's not a broad sort of meta-field or something like that

cin says, "I was telling him that he should participate in our online conference, or try the onsite one"

Tari says, "a lot of it is, i think, a search for a name and job description that will get you those people you envy other schools for having, even though you don't quite understand them"

Tari says, "and don't want to pay for them"

cin says, "and he said 'but I'm not a writing specialist' "

Brooke laughs aloud

Sherri [to cin]: interesting

cin says, "this comes from a guy whose work on the web is all about writing and words...not strictly literature."

Brooke nods to cin

KarenL says, "All--you'd probably be interested in the Internet Researchers Conference; deadline for submissions is Mar 2"

cin [to KarenL]: do you have a URL?

KarenL says, "That's another direction C&W is going in"

Tari says, "there's a position that's usually filled by a grad student: the person has a lot of technical capability and also a lot of experience with and knowledge about teaching with computers"

KarenL says, "just a second"

Brooke [to cin]: do you think he was softpeddling around the topic 'pedagogy' --i.e., he doesn't see himself as someone who hangs out with pedagogy heads?

James nods to Tari

Tari says, "and they do things like help other teachers, run networks, serve as a liason between the strictly techs and the teachers..."

KarenL says, "Here are last year's sessions: http://www.cddc.vt.edu/aoir/2000/schedule.html """

Brooke [to Tari]: . . . and never finish their dissertations . . .

Tari says, "yeah"

Kiwi says, "the person hired here as computers-and-writing specialist doesn't do C&W--she prefers to present at conferences on social justice and feminism; she said C&W "isn't the only conference that 'does' technology." "

cin [to Brooke]: you know, I don't think so -- he's too involved in it here on campus (he's on the technology in the classroom committee, helps grads learn to use the computer classroom, etc)

mday comes in from the hallway.

James waves to mday

mday waves.

cin nods to brooke too much abuse of grads going on still

Kiwi waves to mday.

Tari says, "the problem is that everyone wants one of those, but there's no job title for it because it's happening spontaneously and for free, and no one knows how much the position is worth or that it entails or how to go about making it a legitimate job--AND they don't want to add it to the budget anyhow"

mday had to take a candidate out to dinnner

cin says, "hey mday!"

Brooke [to cin]: pretty puzzling . . .

KarenL says, "Here is this year's Call for Papers: http://www.cddc.vt.edu/aoir/ "

Tari says, "i was that person at my school for a long time"

cin nods, I think he felt that Computers & Writing excluded Literature

Tari says, "and i'm occasionally asked will i do that again"

cin [to KarenL]: thanks

KarenL says, "A lot of the topics for C&W would be appropriate for the Assoc of Internet Researchers"

Sherri [to KarenL]: thanks

James kinda likes UTexas' 'Computers and English Studies' as a grad specialty

cin had the temerity to ask to get paid for work she did over the summer once (running a computer classroom while the professor usually in charge was on sabbatical)

Tari [to cin]: did they do it?

cin says, "the department head told me that I should be glad for the opportunity, and to stop "carrying a cross on my back.""

cin [to Tari]: feh, no

James ugghs

mday says, "ooh, definitely wrong, cin"

Sherri takes cin's cross

Tari [to cin]: i'd have said 'okay, i will--here's your cross back! tell me who to give the root password to!'

cin grins at sherri

James grins at Tari

Sherri [to cin]: it ain't heavy

cin [to Tari]: I said something similar

Brooke switches to Klezmer music on her Internet radio to avoid getting depressed

cin says, "but, the next year, they actually budgeted in funding for me and a few others who were doing work for free (department web site, etc)"

cin says, "it took some time, but they started to listen"

Tari says, "i think a problem is that they don't know how much stuff is worth"

[9:01 pm]

cin nods to tari, it's hard to put a price on knowledge management

cin says, "colleges can't compete with industry wages"

Brooke [to cin]: there was an article in Technical Communication (I think) recently advising tech writers on assessing their knowledge work in accounting terms

Tari [to cin]: wait, what problem is that? that people don't know what the work is worth?

cin [to Tari]: in part

KarenL says, "The TECHR-L list once advised a tech writer to double her prices"

KarenL says, "She did...and none of her clients batted an eye."

Tari notes that the industry generally knows what tech comm and mutimedia are worth these days.

cin says, "but also that what is knowledge management in industry is just sort of intellectual studies for academics."

KarenL says, "In fact, she got more clients."

Sherri says, "bye, all"

KarenL waves bye to Sherri

Tari says, "also, knowledge management is wicked hot, but i think they need to pull in more library scientists"

James waves to Sherri

cin nods to KarenL, when I was consulting, one of my clients (government) said that I wasn't charging enough. He told me to charge more so he could hire me. :)

Sherri has disconnected.

cin waves bye to sherri

cin grins at tari, you're right

cin says, "but...I LOVE TO TEACH! I like working with students. And I like the flexibility in my schedule."

Tari says, "an information management team consisting of some tech comms and some library scientists would be unstoppable"

cin says, "so the lower salary is okay with me"

mday says, "what Cin said, tho"

Tari says, "well, but wait"

cin grins, let me restate that

Tari says, "you have to take all that into consideration if you're thinking about
    salary"

cin says, "I'll put up with the lower salaray...or I chose to take that."

Kiwi nods at cin.

Tari says, "what i'm saying is that if you're being asked to be the humanities
    computing specialist for, say, $40K, you're being asked to meld like three
    professions"

Kiwi is making less now than she was 10 years ago--w/o even a B.A.--but I'm
    enjoying this more (well, most of the time, anyway)

cin [to Tari]: you're right

Tari says, "you're supposed to know the technical side AND the teaching side
    AND be an administrator AND spend your life at work"

KarenL [to cin]: & mday I'm in a class that's studing industry/university rela-
    tions, including consulting work

Kiwi [to Tari]: in the corporate world, I was expected to spend my life at work,
    too.

mday says, "that I no longer do. My dept has a person who does that."

Kiwi shrugs.

cin sighs, my turn to head out I'm afraid. I have a quick meeting (online) with a
    student, then dinner (finally).

cin will see you all next week!

KarenL waves to cin

Brooke waves to cin

CYBORG RESPONSIBILITY

Building on issues raised in the extended excerpt just presented, and thus the
influence of some of the computers and writing community's many voices, I next
turn to what I'll call *cyborg responsibility*. Cyborg responsibility is a unique concept for the computers and writing community, and it is an ideal standard against
which actions may be measured, rather than a road map for such actions. Few
scholars would identify the cyborg figure with any significant responsibility,
whether they're thinking about Case in William Gibson's *Neuromancer* (1984),
the feminist leader Donna Haraway (1985) develops, or even the "cyborg citizen"
Chris Hables Gray (2001) recently introduced. Fewer still would identify any
degree of cyborg responsibility with ties to a particular community, like computers and writing. What cyborg responsibility is, then, is antiresponsibility, or
aresponsibility. To be responsible on cyborg terms is to resist typical explanations
or interpretations of responsibility. Instead, it is to ask critical questions about
issues and experiences in the computers and writing community, no matter the
degree to which an individual is personally invested in anything being questioned.
It is to find value in arguments for and against any project or initiative, rather than

one or the other, and it is to think about who's silent in any conversations, as well as who's speaking. Cyborg responsibility is about creating conflict amid consensus and pushing issues beyond their scope. It promises to open spaces for diversity and inclusiveness, even and perhaps especially when such opening is not easy, and it relies on individuals willing to make professional and personal sacrifices for the betterment of the future. Cyborg responsibility will not breed popularity, however. On the contrary, some individuals in the computers and writing community now, as well as in any future forms it takes, may resent the many questions asked and the additional time taken to make decisions. But, in many respects, the anti- or apopularity matches with the anti- or aresponsibility from which cyborg responsibility emerges. Change agents rarely win popularity contests, of course.

It's important to acknowledge that cyborg responsibility in computers and writing and other communities represents significant risk taking. Any individual who takes this approach will have to account for her or his actions in individual contexts. That said, cyborg responsibility is also a term that needs to be modified or altered, woven into contexts where the risk is most great; this is what I mean when I suggest that the notion is an ideal, rather than a road map or realistic assessment of what specific approaches should be taken in various scenarios. If I am an adjunct or part-time instructor at a university, then I may not be in a position to pursue cyborg responsibility completely, so I'll need to adapt. I might question decisions made about technology and teaching, for instance, but do so in a way that still communicates investment in the department and its future, rather than the more confrontational style the term and its ideal form suggest. Such modifications do not deteriorate the idea of cyborg responsibility because making the modifications requires they be relative to cyborg responsibility, thus reflecting some of the spirit of the ideal concept.

Ultimately, cyborg responsibility, like the agenda in this chapter for the computers and writing community, must reflect the actions of many individuals, rather than one or two. Just as someone, no matter their standing, may not be able to pursue cyborg responsibility as it has already been defined in ideal terms, others will likely not be able to pursue every component of this agenda themselves, whether it's a choice or what their individual contexts require. What makes a difference, however, is having a concept like cyborg responsibility to help direct efforts toward diversity and inclusiveness, when possible. And having an agenda grounded in that responsibility specifically for the computers and writing community should, I hope, lead us toward an equitable and responsible future.

AN AGENDA FOR THE FUTURE OF COMPUTERS AND WRITING

With this introduction to cyborg responsibility and its dimensions, I now offer the four-item agenda. As I previously explained, relative to cyborg responsibility, any agenda like this one will require alteration and careful attention to the contexts in which it is to be introduced. The agenda does, however, provide a beginning, an articulation of important and responsible emphases to be pursued. The spirit of the individual items reflects an emphasis on diversity and inclusiveness at all times.

Item One: Remember Individuals in Any Technology and/or Technology-Adoption Decision

Although seemingly a simple matter, thinking about individuals in technology and/or technology adoption decisions has proven to be less and less evident in recent years. Institutions or organizations sometimes seek the most advanced technologies without any regard for what those technologies might mean for the individuals set to use them, for instance. Or, at the other end of the spectrum, some institutions and organizations do not budget monies for computers and other technologies that individuals could use in important ways. These sorts of decisions are often imagined as cut and dry; that is, either institutions and organizations have monies available, or they do not. But I believe that interpretation masks the importance of individuals in those decisions, a problematic masking against which our cyborg responsibility is to argue.

First I call for more research into any decisions to acquire, use, or otherwise obtain technologies. At the present time, many institutions or organizations make decisions based on either funding available or the features of a technology being considered; the institutions and organizations rarely do more. Discouraging, however, is that many options exist for doing more, but are just not often pursued. An institution or organization might, for instance, set up testing teams to utilize a technology in context, perhaps a teacher's piloting the use of courseware in a course on classical rhetoric or a course on advanced composition or perhaps a sustained and engaged discussion with an institution or organization already making use of the technology. Such opportunities would be promising ways to learn more about individual technologies and their usefulness before a final decision must be made, instead of seeing the decision made and the struggles for implementation afterward. At the same time, such research enables individuals to feel that they have a more active stake in decisions made, a critical participation and investment in any technology's potential for success. Across such predecision research opportunities, our cyborg responsibility is to encourage many different individuals to weigh in and become involved, especially those whose voices are not often enough heard with respect to technologies. When a technology is first being considered, our responsibility is to make certain that the decision makers have pursued appropriate predecision options for learning about the possible choice. It's not enough simply to critique a decision process or encourage predecision research; ultimately, we must act and demonstrate a willingness to do the hard work involved with such a process, as any critique or encouragement we might introduce would necessarily carry less authority if we're not willing to invest ourselves.

Second I call for any acquisition, use or other obtaining of technologies to be determined at least in part specifically and explicitly relative to individuals at all times. To begin, let me demonstrate the difference between technology-centered and individual-centered explorations of a technology's potential to be adopted in classrooms or other settings. The following table models both views for Adobe *Photoshop:*

Photoshop Features Through a Technology-Centered View	Photoshop Features Through an Individual-Centered View
Importing of graphics and images	Enables individuals to create graphics and images
Modification of graphics and images (filters and other special effects)	Enables individuals to modify existing graphics and images
Retouching of graphics and images (especially photographs)	
Painting and drawing	

As readers can see, the rhetorical difference between the two is significant: The first emphasizes what the technology itself can do independent of any context, and the second emphasizes what individuals can do in application. What the rhetoric makes evident, though, is that it's not just the rhetoric itself that is at stake. If a technology is being considered for adoption via its features only—whether reflecting the vision of an institution or organization or instead simply to be able to list the features among those that a specific facility or initiative offers—then that puts individuals at a disadvantage because they have to consider carefully which features prove applicable to their own work. We simply cannot assume that everyone would be able to see immediately the applicability of a technology. In the individual-centered view, however, the possible applicability is clear: If someone believes she or he would like to create or modify graphics and images, then the technology provides an interesting option. If not, then the technology would not be as useful. Our responsibility in computers and writing is to make sure individuals do have such chances to make informed application-based decisions, rather than being forced to scan a list of features and try to determine what might be interesting and useful.

Remembering individuals in decision-making will prove critical to the future, and our responsibility is to facilitate, encourage, and even compel such remembering in any decisions with which we are involved. Not only does this responsibility mean that we should offer ideas and make specific suggestions, but it also illustrates the way we ourselves must become actively involved and model attention to individuals if we expect such attention to be given across institutional or organizational contexts.

Item Two: Actively Seek and Promote Diversity

Our cyborg responsibility is to seek and promote diversity actively, taking professional and personal initiative to involve individuals who do not have a stake in important discussions now, but should. We must recruit the most diverse range of individuals possible to express their views about community issues. Our attention to and inclusion of diversity represents our investment in it, to put the matter simply, so our action is vital to pair with our rhetoric.

First I call for us to emphasize diverse voices in our teaching. In subject areas, we should feature many perspectives on key issues, rather than one or two. If I am teaching a unit on Web site design and usability, for instance, I need to go beyond Jakob Nielsen's (1993, 2000) various perspectives, though they are often regarded as authoritative now, to include alternate views like those of Joann T. Hackos (2002; see also Hackos & Redish, 1998). If students have an opportunity to think about these other views, they will make much more informed decisions about usability, no matter which view they ultimately believe most right. Another way to emphasize diversity in teaching is to explore and negotiate with students the idea of authority. When I teach English composition, I often require students to cite sources they do not believe they would be able to cite in other classes—highly opinionated materials, an 8-year-old child's Web site—because I want them to see that authority is relative, not absolute. This attention to authority helps students to realize that some of the generalizations they had been taught for years— like the idea that books and print journals are innately more valuable to cite than Web sites—fall short at times. Last, we can emphasize diversity in teaching by thinking about *who* we teach, as well as *what* we teach. When I was a faculty member at Furman University, from 1999 to 2001, for instance, I ran a number of computer-based workshops for staff members, and one of the most rewarding initiatives was helping the janitorial staff to access and use e-mail. Sometimes we are in such a hurry to learn about new advanced technologies that we forget about the energy and excitement we had ourselves when we first used e-mail, designed a Web site, or even revised and edited an essay with a word processor. Emphasizing diversity in teaching can help us remember.

Research and publication offer additional opportunities for us to seek and promote diversity. First, we can make diversity the subject of research studies themselves. The Georgia Institute of Technology's Graphics, Design, and Usability Center (GVU) has performed more than 15 surveys of the World Wide Web's users around the world, for instance, and their work successfully documents how the Web is becoming less and less a predominantly Western upper- and middle-class innovation. Scholars around the world are likewise performing research on what has been termed the *digital divide,* looking at who has access to and uses computer technologies in various national, regional, and local contexts (Compaine, 2001, for instance). Another way we can support diversity is by promoting a wide range of publication venues—print and electronic, formal and informal, and static and dynamic. With *Kairos: A Journal of Rhetoric, Technology, and Pedagogy*, which Douglas Eyman and I co-edit, we offer not just an opportunity for scholars to work in hypertextual and other electronic genres, but also advocacy for electronic work to be valued across both traditional and progressive contexts. Our goal is not to promote one or the other per se; rather, it's to emphasize the important contributions they make together, an approach we hope other editors of various print and electronic journals likewise pursue. Last, we can promote diversity by inviting a wide range of scholars at all stages of their professional careers to be involved in projects we lead. Graduate students have proven to be strong contributors to a number of essay collections (e.g., Hawisher & Selfe, 1999; Inman, Reed, & Sands, 2004), and we should expand the scope of such work even further to be sure adjuncts and other part-time faculty have a chance to share their views in them as well.

Emphasizing diversity in the work that we do into the future should help create a future that is more diverse and inclusive. As with other emphases on this agenda, it requires active engagement, rather than simply offering critique or even asking tough questions. We simply must take on professional and personal responsibility for being certain that the future is as diverse and inclusive as possible.

Item Three: Articulate and Model Resistance

For individuals not dedicated to technology and its rise into education, the current era can seem overwhelming, a mess of gigabytes, pixels, and microscopic circuitry that makes no sense whatsoever, but seems quite threatening. In calling for listening to alternate voices, I suggested that we should turn to those who might resist and solicit their opinions. At the same time, however, I think it's a critical part of our cyborg responsibility that even those of us who most strongly advocate technologies and their applications find ways to articulate resistance. Such resistance offers not just another voice to ongoing discussions of the critical implications of technologies, but also a chance to demonstrate careful and thoughtful resistance.

First I call for those who have been strong advocates of technologies to consider how they might resist the advance of such technologies themselves sometimes, helping them understand the point of view or perspective of those who do articulate resistance more often. In some regard, it's simply a matter of being reflective and realistic all at once. Those who use computers and other technologies regularly did not come into life as technological sophisticates; rather learning was gradual, often with parts frustration and elation. Yet many appear to forget this learning curve when they become confident with technologies, articulating frustration that their colleagues, students, or others simply don't "get it." Certainly reductive thinking colors some individuals' minds and the way they come or do not come to technology; if someone sets her or his mind not to invest in technology, even if it appears useful, then that person will not explore options in good faith. At the same time, though, reductive thinking reflects the way those of us comfortable with technology position those who resist. One of the most ironic terms visible in the computers and writing community now for those who resist is *Neo-Luddites,* a reference to the Luddites who resisted industrialization violently in late 19th-century Britain, so when we use the term, we imply not just that those who resist technology do not favor its prominence, but also that they are violent and willing to battle for their view. Such a claim, obviously, falls short of reasonable on almost every occasion. Indeed our cyborg responsibility is to know better by reflexively examining our own resistance, both in the past as we learned about various computers and other technologies we use today, and in the present as we continue to encounter, think about, and deal with new technologies. If we imagine ourselves in the same continuum as those who do not choose to utilize computers and other technologies now, then we create a softer, more appropriate view of them and their positions—one linked to our own development and perspectives over time. The point is not finally that everyone is on a standard continuum that runs from not using computers and other technologies to making use of every possible technology; rather it's that neither extreme is reasonable nor remotely accurate for the circumstances in our contexts. We can use this grounded knowledge to open important interactional spaces about technologies and their impact—spaces with room for resistant perspectives, as well as advocacy.

Second, I call for scholars known for advocating technology to model resistance in publications, presentation, or other venues. Currently what we regard as resistance takes many forms, including off-the-cuff reactions alongside well-reasoned thinking, and this reality is problematic because the two need definitely to be defined and valued differently. We first have to remember, then, that any frustrations we have with colleagues who have never used computers and other technologies, but still critique them anyway, are frustrations they have with us and our typically positive rhetoric about the technologies and their potential to influence our professional and personal lives. The better position is one in the middle, where we acknowledge that everyone will be too resistant at times and too enthusiastic at others and where we help each other ground such responses carefully and thoughtfully. It is in this middle position where we can model resistance effectively. If I am evaluating Web site design applications for my department, for instance, my responsibility should be not just to determine which would be most useful, but also to develop detailed analyses of the individual applications themselves with specific attention to how well a range of users will do with them. In these analyses must be critical questions and areas where I believe the application may not be as useful—perhaps even areas where I think the application will not be effective at all. If I recommend Macromedia's *Dreamweaver* in the end, I cannot simply say that I think it's best, then; instead I have to demonstrate that I have thought through tough critical questions and their implications by modeling the logic I used in responding to the questions and by relating that logic thoroughly in my final recommendation. This sort of detailed analysis demonstrates why decisions are best made after evaluating a technology, why delving thoroughly into a technology does not make one mesmerized into blind adoption, but instead empowers her or him to speak more authoritatively about the positive and negative dimensions of the technology at length. Modeling careful and thoughtful resistance in this fashion demonstrates clearly why both "Wow! This technology changes my world and makes it a happy joy–joy place" and "Technology is Satan in mechanized form" rhetorics fail to meet our needs at this point by representing the sort of reductive thinking that does more harm than good, thus that promises to do more harm than good in the future.

Emphasizing resistance in these ways promises to offer a diverse and inclusive path to the future by inviting everyone to talk about computers and other technologies, whether advocates, resistors, or somewhere in between. More it illustrates commonality among all individuals, helping everyone see their own development and perspectives as linked to those of others. Last, it provides a critical service by modeling careful and thoughtful resistance, rather than quickly or otherwise ill-conceived positions that emerge more from an in-the-moment emotional response rather than well-developed thinking.

Item Four: Participate in the Design of Technologies

More and more, almost all of us find ourselves amid computers and other technologies, whether we seek that position or not. What individuals in the computers and writing community often fail to see, however, and it's an important omission, is that we can and should be contributing to the design of these technologies, rather than acting as passive consumers. Our cyborg responsibility in this case is to enter ourselves into discussions about technology development and application, taking on leadership and advocacy roles as appropriate and making sure as often as possible that community opinions are well represented in any technologies introduced or recrafted.

First I call for everyone to offer feedback about technologies they employ. In this case, I'm talking about going beyond the three-by-five note card response opportunities any purchase enables to providing a more detailed response, whether via a telephone call, e-mail, a Web site, or even a formal letter. In doing so, we should be mindful of our rhetoric, as well as the investment that technology designers have in their various innovations. Just as we would not want to be told that our teaching approach or scholarly writing was poor and ill-conceived, so it is not appropriate to interact with the technology designers by using that sort of rhetoric. "Errors" and "mistakes" can be framed as "areas of concern" or "suggestions for improvement" and can be delineated amid an overall rhetorical approach that also identifies strengths of a technology, thus reflecting a more generous and equitable, yet still concerned approach. If we adopt such a careful rhetoric and if we take on our responsibility to provide feedback to technology designers, then we've acted proactively for the future, as well as demonstrated our willingness to act as well as critique.

Second I call for everyone to learn about and become invested in the design process itself, including early discussions about the sorts of technologies that could best support their work, to usability and other testing of prototypes, and to ongoing discussion about future versions. Taking this approach positions us not just as respondents to particular technologies introduced to us, but also as active agents in the creation and revision of those technologies, and this positioning is a critical component of our cyborg responsibility. In these discussions, we will have many chances to argue for diversity and inclusiveness, and we can help technologies be all the more attentive to individuals and their various contexts. To put it simply, we have valuable expertise and experience as individuals invested in computers and writing, no matter the contexts in which we are working, so we can make a difference by carefully articulating the goals and responsibilities of our various contexts and suggesting ways a technology might be more useful in them. If a reader with considerable teaching experience has a technology in mind that has not yet been created, but would make her or him a more effective teacher by enabling new teaching and learning approaches to be introduced, then that thought could even be the genesis of a technology development project. That is, we do not have to think only about technologies that have been suggested to us, whether in the beginning stages of development or those later; instead we can come up with our own ideas and pitch them to technology developers as innovations they should consider. Either way, we should take advantage of our opportunity to be involved and offer insight to designers. It's a chance for us to do more than simply critique technologies and talk about how designers need to know more about our contexts, after all. Plus, we stand to learn a great deal from the designers and from the involvement we have with technology design and development. Many scholars have written about how teaching in a new environment helped them become better teachers not just in the new environment, but also ones they have occupied for years, and the same logic would be evident here. Being involved with the design and development processes around technologies will help us become better in our individual contexts by forcing us to think carefully about what we do and what we value, as well as what supporting technologies might and might not contribute.

With this investment in the design and development of technologies, we can contribute a great deal to any future involving such technologies, no matter how it unfolds, and we have positioned ourselves as participants, practitioners willing to take on additional responsibility for the betterment of our institutions and organizations, as well as ourselves.

IMPLEMENTING THE AGENDA

Implementing an agenda like the one just mentioned requires a considerable investment of time and energy and a realistic sense of how such implementation may proceed. One scholar who has studied implementation carefully is Everett Rogers (1995). In *Diffusion of Innovations,* he explores how any *innovation,* a term he defines as ". . . an idea, practice, or object that is perceived as new by an individual or other unit of adoption" (p. 11), enters systems, including how the innovation is ultimately adopted or rejected, as well as how the innovation changes itself and changes the systems into which it is introduced. The innovation in our case is the aforementioned agenda. As explained by Rogers (1995), *diffusion* is the entire process: "Diffusion is the process by which an innovation is communicated through certain channels over time among the members of a social system" (p. 5). Because of diffusion's emphasis on the broad term innovation and its interdisciplinary, time-tested character, it proves especially useful here as a way of thinking about how this chapter's agenda might be implemented in and across various computers and writing community contexts.

One of the first lessons we need to take from diffusion research is that change takes time—sometimes a great deal of time. When thinking about adopting this chapter's agenda then, or more generally working to be responsible in cyborg terms, we cannot grow impatient. Rogers suggests that the adoption of any innovation proceeds along an *s*-curve, as shown in the following diagram.

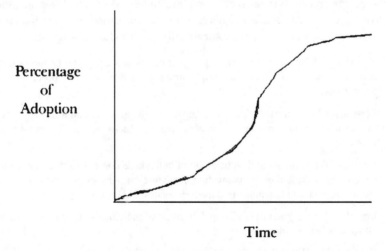

In the diagram, the horizontal axis represents time, so we might think about it in terms of weeks or days, depending on what seems most appropriate for the contexts in which we're working. The vertical axis, on the other hand, represents the percentage of individuals in any system who have chosen to adopt an innovation. In our terms, we're thinking about percentage of individuals in a computers and writing community context that have considered and adopted an item of the agenda, or item adapted from the agenda. What's most important about the s-curve diagram is how it represents adoption stages. Readers should notice three specific stages, in fact. In the first stage, only a small percentage of individuals adopts an innovation; these are what Rogers would term *opinion leaders* or *change agents* and who we know as colleagues who push boundaries, asking tough questions and thinking creatively. In the middle stage, reflected by the steeper vertical trajectory of the *s,* many individuals adopt an innovation, following the lead of the change agents, but not themselves taking tremendous risks. And, finally, in the last stage, where vertical trajectory flattens again, we have what Rogers (1995) would term *laggards,* who are individuals who are among the last to adopt an innovation. As we work toward implementing items of this chapter's agenda, we should be mindful of all of these stages. Clearly we cannot expect a large number of individuals to rush to the innovation, but we can gain the sense that if change agents pursue elements of the agenda, those elements have a good chance of being adopted more widely. After any strong increase in support for an agenda item, we also must expect continued resistance, leading ultimately perhaps to late adoption, at best.

Diffusion also teaches us how the adoption process actually proceeds, articulating stages that we can use to measure the progress of our agenda items and other initiatives. It would be reductive at best, after all, to imagine that an agenda item will either be adopted or not and thus that the process comes down to a single vote or choice; the process is much more complex. Rogers (1995) identifies the following five stages, which I'll adapt to make them more specific for the computers and writing community and even more specifically for this chapter's agenda:

- Knowledge: When an individual or group of individuals encounter an innovation and understands its character and implications, as well as operating procedure, if appropriate;

- Persuasion: When the individual or group of individuals makes an informal determination about the innovation, whether positive, negative, or somewhere in between;

- Decision: When the individual or group of individuals comes to a decision about the possible adoption of the innovation, including formally expressing that opinion in a response to the innovation itself in whatever context is appropriate;

- Implementation: When the individual or group of individuals puts the innovation into action, quite simply; and

- Confirmation: When the individual or group of individuals reassesses their adoption decision, determining that the innovation is indeed appropriate, needs changes in order to be appropriate, or should not be further utilized. (p. 162)

Together these five stages offer a large-scale look at how an innovation is considered. Where the five-step process becomes most valuable as a model, however, is when it is superimposed onto the s-curve shared earlier, as that enables us to see that much work should be done before even the knowledge stage of the process can be confirmed. Certainly a single individual can be exposed to an innovation, like this chapter's agenda, and that reflects some degree of knowledge, but if we're talking about larger contexts, like departments or even institutions, then one individual's exposure is a long way from a critical mass of exposure such that a group decision could be made. In the computers and writing community, we operate often in just such larger contexts, so we can anticipate the adoption process for this chapter's agenda and any other agendas to be long and involved, requiring promotion from opinion leaders and change agents initially and then careful thinking by a group later, thinking that will inevitably result in modifications.

Last, diffusion theory includes explicit attention to *implementation,* helping us think in great detail through the way agenda items we pursue may or may not be carried out successfully. Rogers, in particular, emphasizes the role of *re-invention,* a concept perfectly matching this chapter's argument that agenda items should be modified for various computers and writing contexts and that cyborg responsibility is an ideal that should be adapted into specific contexts, rather than taken as an absolute or fixed standard. At the same time, Rogers (1995) points to the critical implications of reinvention as a component of implementation. If someone suggests an agenda item reflecting a modification of one of those presented in this chapter or a completely new agenda item, then she or he has a significant stake in the suggestion such that reinvention may be considered an affront, an implied or even explicit indication that the suggestion wasn't appropriate or reasonable. Clearly such a critical perspective has merit. We rarely find an innovation in the computers and writing community that persists just as it was introduced, after all, and feelings could emerge, if the reinvention isn't framed as collaborative and democratic. In this sense, diffusion theory's emphasis on reinvention reminds us to be careful how we present suggestions and how we otherwise build on ideas introduced.

With these perspectives from diffusion theory, and particularly the scholarship of Rogers (1995), we can develop recommendations for implementing the agenda proposed in this chapter, as well as other agendas reflecting cyborg responsibility that are developed in the future for the computers and writing community. Specific recommendations include the following:

- Allow plenty of time: The adoption process may involve weeks, months, or even years, depending on the nature of the innovation and especially for large-scale changes like those suggested by this chapter's agenda;
- Understand the constituencies at work and the need for diverse voices: With innovations like the agenda articulated in this chapter, many different individuals will be affected, including those across institutions and organizations in the computers and writing community, so developing the innovation with attention to this diversity offers a more promising option;
- Anticipate reinvention and avoid being defensive: While it's important to have a stake in any innovations being developed, that stake should not breed defensiveness, especially if it's associated with an innovation like this chapter's agenda that promises to impact so many different individuals;

- Regard any decisions as subject to future considerations, rather than thinking of them as permanent: Adoption of an innovation is certainly an exciting
  outcome, if the innovation promises to contribute to the computers and writing community or other contexts, but it should not be seen as a complete
  endorsement of the innovation because the innovation itself may be
  required to change or may even be seen as not effective.

In addition to thinking about and assessing the implementation process for any
innovation like this chapter's agenda, readers should also consider sharing the
details of that process in professional forums, like presentations and publications,
if the process can shed new light on how innovations proceed in the computers
and writing community. Not only would such reflective thinking benefit the project and the individuals involved, but it may also enable community members in
other contexts to take advantage of lessons learned.

CONCLUSION

The agenda developed in this chapter serves as a endpoint for *Computers and
Writing: The Cyborg Era,* but also a starting point for where the community itself
might go. The goal, ultimately, is less about pursuing a specific future and more
about how the community continues to develop: if its activity prominently features a diverse range of voices, if these voices reflect an emphasis on inclusiveness
in the community, and more. Whatever today's computers and writing community
becomes tomorrow, whether a transformed single community or many new ones,
the emphases in this chapter's agenda, as well as the ideal concept of *cyborg
responsibility,* should help the future to be equitable and responsible.

References

Allen, N. (1996). Gaining electronic literacy: Workplace simulations in the classroom. In P. Sullivan & J. Dautermann (Eds.), *Electronic literacies in the workplace: Technologies of writing* (pp. 216–237). Urbana, IL: National Council of Teachers of English.

Alliance for Computers and Writing. (2001). *Home page.* Retrieved November 1, 2001, from http://english.ttu.edu/acw

Arp, L. (1990). Information literacy or bibliographic instruction: Semantics or philosophy? *Reference Quarterly, 30,* 46–9.

Association of College and Research Libraries. (2000). *Information literacy competency standards for higher education.* Retrieved November 2, 2001, from http://www.ala.org/acrl/ilcomstan.html

Balsamo, A. (1996). *Technologies of the gendered body: Reading cyborg women.* Durham, NC: Duke University Press.

Barker, T., & Kemp, F. O. (1990). Network theory: A postmodern pedagogy for the writing classroom. In C. Handa (Ed.), *Computers and community: Teaching composition in the twenty-first century* (pp. 1–27). Portsmouth, NH: Boynton/Cook.

Baron, D. (1999). From pencils to pixels: The stages of literacy technologies. In G. E. Hawisher & C. L. Selfe (Eds.), *Passions, pedagogies, and the 21st century technologies* (pp. 15–33). Logan: Utah State University Press.

Behar, R. (1996). *The vulnerable observer: Anthropology that breaks your heart.* Boston, MA: Beacon.

Behrens, S. (1994). A conceptual analysis and historical overview of information literacy. *College and Research Libraries, 55,* 309–22.

Birkerts, S. (1994). *The Gutenberg elegies: The fate of reading in an electronic age.* Boston, MA: Faber & Faber.

Bishop, W. (1990). *Something old, something new: College writing teachers and classroom change.* Carbondale: Southern Illinois University Press.

Bishop, W. (1999). *Ethnographic writing research: Writing it down, writing it up, and reading it.* Portsmouth, NH: Heinemann.

Bitter, G. (1986). Computer literacy: Awareness, application, and programming. New York: Addison-Wesley.

Blair, K., & Takayoshi, P. (Eds.). (1999). *Feminist cyberscapes: Mapping gendered academic spaces.* Norwood, NJ: Ablex.

Boese, C. (1998). *The ballad of the internet nutball: Chaining rhetorical visions from the margins of the margins to the mainstream of the Xenaverse.* Unpublished doctoral dissertation, Rensselaer Polytechnic Institute, Troy, NY.

Bolter, J. D. (1992). Literature in the electronic writing space. In M. Tuman (Ed.), *Literacy online: The promise (and peril) of reading and writing with computers* (pp. 19–42). Pittsburgh: University of Pittsburgh Press.

Briggs, C. L. (1986). *Learning how to ask: A sociolinguistic appraisal of the role of the interview in social science research.* New York: Cambridge University Press.

Brooks, G. (1995). *Nine parts of desire: The hidden world of Islamic women.* New York: Anchor.

Bruce, B., & Hogan, M. P. (1998). The disappearance of technology: Toward an ecological model of literacy. In D. Reinking, M. McKenna, L. Labbo, & R. Kieffer (Eds.), *Handbook of literacy and technology: transformations in a post-typographic world* (pp. 269–281). Mahwah, NJ: Lawrence Erlbaum Associates.

Bruce, C. (1997). *The seven faces of information literacy.* Adelaide, Australia: Auslib Press.

Bruckman, A. (1993). Gender swapping on the internet. Retrieved November 3, 2001, from ftp:// ftp.cc.gatech.edu/pub/people/asb/papers/gender-swapping.txt

Bruckman, A. (1997). *MOOSE crossing: Construction, community, and learning in a networked virtual world for kids.* Retrieved November 12, 1999, from http://www.cc.gatech.edu/~asb/thesis/ index.html

Bruner, C. (1999). *The new media literacy handbook: An educator's guide to bringing new media into the classroom.* New York: Anchor.

Bush, V. (1945, July). As we may think. *Atlantic Monthly, 176*(1), 101–108.

Butler, W. (2001). Where we come from. *Kairos: A Journal for Teachers of Writing in Webbed Environments, 6*(2). Retrieved December 15, 2001, from http://english.ttu.edu/kairos/6.2/features/ townhalls/butler.htm

Chapman, D. W. (1999). A Luddite in cyberland, or how to avoid being snared by the web. *Computers and composition, 16*(2), 247–252.

Clinton, W. J. (1998, June 5). Remarks by president at Massachusetts Institute of Technology 1998 commencement. Retrieved November 3, 2001, from http://clinton2.nara.gov/WH/New/html/ 19980605-28045.html

Clynes, M., & Kline, N. S. (1960, September). Cyborgs and space. *Astronautics,* pp. 26–27, 74–75.

Compaine, B. M. (Ed.). (2001). *The digital divide: Facing a crisis or creating a myth?* Cambridge, MA: MIT Press.

Condon, W. (2000). It's STILL not far to the frontier: Encouraging students to become active professionals in C&W. *Kairos: A Journal for Teachers of Writing in Webbed Environments, 5*(2). Retrieved November 5, 2001, from http://english.ttu.edu/kairos/5.2

Covino, W. A. (1998). Cyberpunk literacy; or, piety in the sky. In T. W. Taylor & I. Ward (Eds.), *Literacy theory in the age of the Internet* (pp. 34–46). New York: Columbia University Press.

Craig, T., Harris, L., & Smith, R. (1998). Rhetoric of the "contract zone": Composition on the front lines. In T. Taylor & I. Ward (Eds.), *Literacy theory in the age of the Internet* (pp. 122–145). New York: Columbia University Press.

Crump, E. (1996). *Interversity: Convergence and transformation, or discovering the revolution that already happened.* Keynote address to the 1996 Teaching in the Community College Online Conference.

Crump, E. (1998). At home in the MUD: Writing centers learn to wallow. In C. Haynes & J. R. Holmevik (Eds.), *High wired: On the design, use, and theory of educational MOOs* (pp. 177–91). Ann Arbor, MI: University of Michigan Press.

Cuban, L. (1986). *Teachers and machines: The classroom use of technology since 1920.* New York: Teachers College Press.

Cummins, G. S. (2000). Centering in the distance: Writing centers, inquiry, and technology. In J. A. Inman & D. N. Sewell (Eds.), *Taking flight with OWLs: Examining electronic writing center work* (pp. 203–210). Mahwah, NJ: Lawrence Erlbaum Associates.

Davis, B. H., & Brewer, J. (1997). *Electronic discourse: Linguistic individuals in virtual space.* Albany: State University of New York Press.

Davis-Floyd, R., & Dumit, J. (Eds.). (1998). *Cyborg babies: From techno-sex to techno-tots.* New York: Routledge.

Doheny-Farina, S. (1996). *The wired neighborhood.* New Haven, CT: Yale University Press.

Downey, G., Dumit, J., & Williams, S. (1995). Cyborg anthropology. *Cultural Anthropology, 10,* 264–269.

Doyle, R. (2000). Uploading anticipation, becoming-silicon. *JAC: A Journal of Composition Theory, 20*(4).

Eldred, J. C. (1999). Technology's strange, familiar voices. In G. E. Hawisher & C. L. Selfe (Eds.), *Passions, pedagogies, and 21st century technologies* (pp. 387–98). Logan: Utah State University Press.

Ely, D. P. (1963). *The changing role of the audiovisual process in education: A definition and glossary of related terms* [TCP Monograph No. 1]. *AV Communication Review, 11*(1), Suppl. No. 6.

Evertts, E. L. (Ed.). (1972). *English and reading in a changing world.* Urbana, IL: National Council of Teachers of English.

Fanderclai, T. L. (1995). MUDs in education: New environments, new pedagogies. *Computer-Mediated Communication Magazine, 2*(1), 8.

Fanderclai, T. L. (2001). Changes in latitudes, changes in attitudes. *Kairos: A Journal for Teachers of Writing in Webbed Environments, 6*(2). Retrieved December 14, 2001, from http://english.ttu.edu/kairos/6.2/features/townhalls/fanderclai.htm

Feenberg, A. (1999). *From essentialism to constructivism: Philosophy of technology at the crossroads.* Retrieved May 1, 1999, from http://www-rohan.sdsu.edu/faculty/feenberg/talk4.html

Fitzsimmons-Hunter, P., & Moran, C. (1998). Writing teachers, schools, access, and change. In T. W. Taylor & I. Ward (Eds.), *Literacy theory in the age of the Internet* (pp. 158–69). New York: Columbia University Press.

Forsythe, A. I. (1969). *Computer science: A first course.* New York: Wiley.

Forsythe, A. I., & Organick, E. I. (1978). *Programming language structures.* New York: Academic.

Freidan, B. (1963). *The feminine mystique.* New York: Laurel.

Freire, P. (1970). *Pedagogy of the oppressed* (M. B. Ramos, Trans.). New York: Herder & Herder.

Geertz, C. (1973). *The interpretation of cultures: Selected essays.* New York: Basic.

Gerrard, L. (1993). Computers and composition: Rethinking our values. *Computers and Composition, 9*(2), pp. 23–34.

Gerrard, L. (1996). Preface. In *Computers and the teaching of writing in American higher education, 1979–1994: A history* (pp. ix–xii). Norwood, NJ: Ablex.

Gerrard, L. (1998). Statement. *Kairos: A Journal for Teachers of Writing in Webbed Environments, 3*(2). Retrieved November 5, 2001, from http://english.ttu.edu/kairos/3.2/coverweb/statements/gerrard.html

Gerrard, L. (2000). "Diets suck!" and other tales of women's bodies on the Web. *Kairos: A Journal for Teachers of Writing in Webbed Environments, 5*(2). Retrieved March 4, 2001, from http://english.ttu.edu/kairos/5.2/binder.html?coverweb/gerrard

Gibson, W. (1984). *Neuromancer.* New York: Ace.

Gillespie, P., & Lerner, N. (2000). *The Allyn and Bacon guide to peer tutoring.* Boston: Allyn & Bacon.

Gilster, P. (1997). *Digital literacy.* New York: Wiley.

Gomez, M. L. (1991). The equitable teaching of composition. In G. E. Hawisher & C. L. Selfe (Eds.), *Evolving perspectives on computers and composition studies: Questions for the 1990s* (pp. 318–335). Urbana, IL: National Council of Teachers of English.

Granville, E. B. (1989). My life as a mathemetician. *Sage: A Scholarly Journal of Black Women, 6*(4), 44–46.

Graphics, Visualization, and Usability Center, Georgia Institute of Technology. (1999). GVU's 9th WWW User Survey. Retrieved May 1, 1999, from http://www.cc.gatech.edu/gvu/user_surveys/survey-1998-04.

Gray, C. H. (2001). *Cyborg citizen: Politics in the posthuman age.* New York: Routledge.

Greco, D. (1995). *Cyborg: Engineering the body electric* [CD-ROM]. Watertown, MA: Eastgate Systems.

Gruber, S. (Ed.). (2000). *Weaving a virtual web: Practical approaches to new information technologies.* Urbana, IL: National Council of Teachers of English.

Gunkel, D. J. (2000). Hacking cyberspace. *JAC: A journal of composition theory, 20*(4).

Haas, C. (1989). Seeing it on the screen isn't really seeing it: Computer writers' reading problems. In G. E. Hawisher & C. L. Selfe (Eds.), *Critical perspectives on computers and composition studies* (pp. 16–29). New York: Teacher's College.

Haas, C. (1996). *Writing technology: Studies on the materiality of literacy.* Mahwah, NJ: Lawrence Erlbaum Associates.

Hackos, J. T, & Redish, J. C. (1998). *User and task analysis for interface design.* New York: Wiley.

Hackos, J. T. (2002). *Managing content for dynamic Web delivery.* New York: Wiley.

Hakken, D. (1993). *Computing myths, class realities: An ethnography of technology and working people in Sheffield, England.* Boulder, CO: Westview.

Hakken, D. (1999). *Cyborgs@cyberspace? An ethnographer looks to the future.* New York: Routledge.

Handa, C. (Ed.). (1990). *Computers and community: Teaching composition in the twenty-first century.* Portsmouth, NH: Boynton/Cook.

Haraway, D. (1985). Manifesto for cyborgs: Science, technology, and socialist feminism in the 1980s. *Socialist Review, 80,* 65–108.

Harste, J., Woodward, V., & Burke, C. (1984). *Language stories and literacy lessons.* Portsmouth, NH: Heinemann.

Hawisher, G. E. (1986). Studies in word processing. *Computers and Composition, 4*(1), 6–31.

Hawisher, G. E., & LeBlanc, P. (Eds.). (1992). *Re-imagining computers and composition: Teaching and research in the virtual age.* Portsmouth, NH: Boynton/Cook.

Hawisher, G. E., LeBlanc, P., Moran, C., & Selfe, C. L. (1996). *Computers and the teaching of writing in American higher education, 1979–1994: A history.* Norwood, NJ: Ablex.

Hawisher, G. E., & Moran, C. (1993). Electronic mail and the writing instructor. *College English, 55,* 627–643.

Hawisher, G. E., & Selfe, C. L. (Eds.). (1999). *Passions, pedagogies, and 21st-century technologies.* Logan, Utah State University Press.

Hawisher, G. E., & Selfe, C. L. (Eds.). (2000). *Global literacies and the World Wide Web.* New York: Routledge.

Hayden, M. (1989). What is technological literacy? *Bulletin of Science, Technology, and Society, 119,* 220–233.

Hayles, N. K. (1999). *How we became posthuman: Virtual bodies in cybernetics, literature, and informatics.* Chicago: University of Chicago Press.

Haynes, C. (1998). prosthetic_rhetorics@writing.loss.technology. In T. W. Taylor & I. Ward (Eds.), *Literacy theory in the age of the internet* (pp. 79–92). New York: Columbia University Press.

Haynes, C., & Holmevik, J. R. (Eds.). (1998a). *High wired: On the design, use, and theory of educational MOOs.* Ann Arbor: University of Michigan Press.

Haynes, C., & Holmevik, J. R. (1998b). Position statement. *Kairos: A Journal for Teachers of Writing in Webbed Environments, 3*(2). Retrieved March 17, 2000, from http://english.ttu.edu/kairos/3.2/coverweb/statements/haynes.html

Heba, G. (1997). Hyperrhetoric: Multimedia, literacy, and the future of composition. *Computers and Composition, 14*(1), 19–44.

Heidegger, M. (1977). *The question concerning technology and other essays* (W. Lovitt, Trans.). New York: HarperCollins.

Herbert, F. (1965). *Dune.* Philadelphia: Chilton.

Herring, S. C. (1993). Gender and democracy in computer-mediated communication. *Electronic Jour-*

nal of Communication, 3(2). Retrieved November 4, 2001, from http://www.cios.org/www/ejc/v3n293.htm

Herring, S. C. (1996). Posting in a different voice: Gender and ethics in computer-mediated communication. In C. Ess (Ed.), *Philosophical perspectives on computer-mediated communication* (pp. 115–145). Albany: State University of New York Press.

Herring, S. C. (1999). Bringing familiar baggage to the new frontier: Gender differences in computer-mediated communication. In V. Vitanza (Ed.), *CyberReader* (pp. 144–154). Needham Heights, MA: Allyn & Bacon.

Hodge, R., & Kress, G. (1988). *Social semiotics.* Ithaca, NY: Cornell University Press.

Hofstetter, F. T. (2000). *Internet literacy* (2nd ed.). Boston, MA: McGraw-Hill.

Hofstetter, F. T. (2001). *Multimedia literacy* (3rd ed.). Boston, MA: McGraw-Hill.

Holdstein, D., & C. L. Selfe (Eds.). (1990). *Computers and writing: Theory, research, practice.* New York: Modern Language Association.

Holt, P., & Williams, N. (Eds.). (1992). *Computers and writing: State of the art.* Boston, MA: Kluwer.

Howard, T., & Benson, C. (Eds.). (1999). *Electronic networks: Crossing boundaries / creating communities.* Portsmouth, NH: Boynton/Cook.

Hutchings, P., & Wutzdorff, A. (1988). Experiential learning across the curriculum: Assumptions and principles. In *Knowing and doing: Learning through experience* (pp. 3–27). New York: Jossey-Bass.

Inman, J. A. (2000). The importance of innovation: Diffusion theory and technological progress in writing centers. *The Writing Center Journal, 21*(1), 49–66.

Inman, J. A., Reed, C., & Sands, P. (Eds.). (2004). *Electronic collaboration in the humanities: Issues and options.* Mahwah, NJ: Lawrence Erlbaum Associates.

Inman, J. A., & Sewell, D. N. (Eds.). (2000). *Taking flight with OWLs: Examining electronic writing center work.* Mahwah, NJ: Lawrence Erlbaum Associates.

Johnson-Eilola, J. (1997). *Nostalgic angels: Rearticulating hypertext writing.* Norwood, NJ: Ablex.

Jordan-Henley, J., & Maid, B. M. (1995a). MOOving along the information superhighway: Writing centers in cyberspace. *Writing Lab Newsletter, 19*(5), 1–4.

Jordan-Henley, J., & Maid, B. M. (1995b). Student impact and college/university collaboration. *Computers and Composition, 12*(2), 211–218.

Jordan-Henley, J., & Maid, B. M. (1995c). Tutoring in cyberspace: Student impact and college/university collaboration. *Computers and Composition, 12*(2), 211.

Jordan-Henley, J., & Maid, B. M. (2000). Advice to the linelorn: Crossing state borders and the politics of cyberspace. In J. A. Inman & D. N. Sewell (Eds.), *Taking flight with OWLs: Examining electronic writing center work* (pp. 105–116). Mahwah, NJ: Lawrence Erlbaum Associates.

Joyce, M. (1995). *Of two minds: Hypertext pedagogy and poetics.* Ann Arbor: University of Michigan Press.

Joyce, M. (1999). Beyond next before you once again: Respossessing and renewing electronic culture. In G. E. Hawisher & C. L. Selfe (Eds.), *Passions, pedagogies, and 21st century technologies* (pp. 399–417). Logan: Utah State University Press.

Kamil, M. L., & Lane, D. M. (1998). Researching the relation between technology and literacy: An agenda for the 21st century. In D. Reinking, M. C. McKenna, L. D. Labbo, & R. D. Kieffer (Eds.), *Handbook of literacy and technology: Transformations in a post-typographic world* (pp. 323–341). Mahwah, NJ: Lawrence Erlbaum Associates.

Keenan, C. (1998). Position statement. *Kairos: A Journal for Teachers of Writing in Webbed Environments, 3*(2). Retrieved June 23, 2000, from http://english.ttu.edu/kairos/3.2/coverweb/statements/keenan.html

Kellner, D. (1964). Introduction. In *One-dimensional man: Studies in ideology of advanced industrial society* (2nd ed., pp. xi–xxxix). Boston, MA: Beacon.

King, M. L. (1964). *Why we can't wait.* New York: Harper & Row.

Knoblauch, C., & Brannon, L. (1993). *Critical teaching and the idea of literacy.* Portsmouth, NH: Boynton/Cook.

Kolko, B. (1998). We are not just (electronic) words: Learning the literacies of culture, body, and politics. In T. W. Taylor & I. Ward (Eds.), *Literacy theory in the age of the Internet* (pp. 61–78). New York: Columbia University Press.

Kress, G. (1999). English at the crossroads: Rethinking curricula of communication in the context of the turn to the visual. In G. E. Hawisher & C. L. Selfe (Eds.), *Passions, pedagogies and 21st century technologies* (pp. 66–88). Logan: Utah State University Press.

Lancaster, R. N. (1992). *Life is hard: Machismo, danger, and the intimacy of power in Nicaragua.* Berkeley: University of California Press.

Lang, S. (2000). Reconciling local and global (or, short term and long term) graduate student concerns. *Kairos: A Journal for Teachers of Writing in Webbed Environments, 5*(2). Retrieved November 5, 2001, from http://english.ttu.edu/kairos/5.2

Lankshear, C., Bigum, C., Durrant, C., Green, B., Honan, E., Morgan, W., Murray, J., Snyder, I., & Wild, M. (1997). *Digital rhetorics: Literacies and technologies in education: Current practices and future directions.* Department of Employment, Education, Training and Youth Affairs, Canberra, Australia.

Lankshear, C., Snyder, I., with Green, B. (2000). *Teachers and technoliteracy: Managing literacy, technology and learning in schools.* St Leonards, Sydney, Australia: Allen & Unwin.

Latour, B. (1993). *We have never been modern* (C. Porter, Trans.). New York: Harvester Wheatstaff.

LeBlanc, P. (1998). Statement. *Kairos: A Journal for Teachers of Writing in Webbed Environments, 3*(2). Retrieved November 5, 2001, from http://english.ttu.edu/kairos/3.2/coverweb/statements/leblanc.html

Lemke, J. L. (1998). Metamedia literacy: Transforming meanings and media. In D. Reinking, M. C. McKenna, L. D. Labbo, & R. D. Kieffer (Eds.), *Handbook of literacy and technology: Transformations in a post-typographic world* (pp. 283–301). Mahwah, NJ: Lawrence Erlbaum Associates.

Lester, P. M. (1995). Digital literacy: Visual communication and computer images. *Computer Graphics, 29,* 25–27.

Lewis, T., & Gagel, C. (1992). Technological literacy: A critical analysis. *Journal of Curriculum Studies, 24*(2), 117–138.

Logsdon, J. M. (1970). *The decision to go to the moon: Project Apollo and the national interest.* Cambridge, MA: MIT Press.

Lombardi, J. V. (1983). *Computer literacy: The basic concepts and language.* Bloomington: Indiana University Press.

Luehrmann, A. (1984). *Computer literacy: A hands-on approach.* New York: McGraw-Hill.

Lumsdaine, A. A., & Glaser, R. (1960). *Teaching machines and programmed learning: A source book.* Washington, DC: Department of Audio-Visual Instruction, National Education Association.

Lyotard, J. F. (1984). *The postmodern condition: A report on knowledge* (G Massumi & B. Massumi, Trans.). Minneapolis: University of Minnesota Press.

Marcuse, H. (1964). *One-dimensional man: Studies in the ideology of advanced industrial society.* Boston, MA: Beacon.

Masat, F. E. (1981). *Computer literacy in higher education.* Washington: American Association for Higher Education.

Mead, M. (1932). *The changing culture of an Indian tribe.* New York: Columbia University Press.

Mead, M. (1933). *Coming of age in Samoa: A psychological study of primitive youth for Western civilization.* New York: Blue Ribbon.

Mikulecky, L., & Kirkley, J. R. (1998). Changing workplaces, changing classes: The new role of technology in workplace literacy. In D. Reinking, M. McKenna, L. D. Labbo, & R. Kieffer (Eds.), *Handbook of literacy and technology: Transformations in a post-typographic world* (pp. 303–320). Mahwah, NJ: Lawrence Erlbaum Associates.

Monroe, B. J., Rickly, R., Condon, W., & Butler, W. (2000).The near and distant futures of OWL and the writing center. In J. A. Inman & D. N. Sewell (Eds.), *Taking flight with OWLs: Examining electronic writing center work* (pp. 211–222). Mahwah, NJ: Lawrence Erlbaum Associates.

Moran, C. (1999). Access: The a-word in technology studies. In G. E. Hawisher & C. L. Selfe (Eds.), *Passions, pedagogies, and 21st century technologies* (pp. 205–220). Logan: Utah State University Press.

Murray, J. H. (1997). *Hamlet on the holodeck: The future of narrative in cyberspace.* New York: Free Press.

Nardi, B., & O'Day, V. (1999). *Information ecologies: Using technology with heart.* Cambridge, MA: MIT Press.

The Netoric Project. (2001). *Netoric home.* Retrieved November 8, 2001, from http://nova.bsuvc.bsu .edu/%7Egsiering/netoric

Nielsen, J. (1993). *Usability engineering.* Boston, MA: Academic.

Nielsen, J. (2000). *Designing Web usability.* Indianapolis: New Riders.

Ohmann, R. (1985). Literacy, technology, and monopoly capitalism. *College English, 47,* 675–89.

Ong, W. J. (1982). *Orality and literacy: The technologizing of the word.* New York: Methuen.

Porter, J. E. (1998). *Rhetoric ethics and internetworked writing.* Norwood, NJ: Ablex.

Potter, W. J. (2001). *Media literacy.* Thousand Oaks, CA: Sage.

Rader, H. B. (1995). Information literacy and the undergraduate curriculum. *Library Trends, 44*(2), 270–8.

Regan, A. E., & Zuern, J. D. (2000). Community-service learning and computer-mediated advanced composition: The going to class, getting online, and giving back project. *Computers and Composition, 17,* 177–195.

Reinking, D., McKenna, M. C., Labbo, L. D., & Kieffer, R. D. (Eds.). (1998). *Handbook of literacy and technology: Transformations in a post-typographic world.* Mahwah, NJ: Lawrence Erlbaum Associates.

Reiss, D., Selfe, D., & Young, A. (Eds.). (1998). *Electronic communication across the curriculum.* Urbana, IL: National Council of Teachers of English.

Rice, R. (2001). E-literacy and orality: The hands-free, voice-activated, any-to-any future classroom. *Kairos: A Journal for Teachers of Writing in Webbed Environments, 6*(2). Retrieved December 10, 2001, from http://english.ttu.edu/kairos/6.2/features/townhalls/rice1.htm

Richardson, S. (1794). *The history of Pamela; or, virtue rewarded.* Worcester, MA: Isaiah Thomas.

Rickly, R. J. (1999). Promotion, tenure, and technology: Do we get what we deserve? In T. Howard & C. Benson (Eds.), *Electronic networks: Crossing boundaries/creating communities* (225–239). Portsmouth, NH: Heinemann-Boynton Cook.

Rodrigues, D. (2001). Statement. *Kairos: A Journal for Teachers of Writing in Webbed Environments, 6*(2). Retrieved December 5, 2001, from http://english.ttu.edu/kairos/6.2

Rogers, E. M. (1995). *Diffusion of innovations* (4th ed.). New York: Free Press. (Original work published 1962)

Romano, S. (1993). The egalitarianism narrative: Whose story? Which yardstick? *Computers and Composition, 10*(3), 5–28.

Rosenberg, M. (1999). Statement. *Academic.Writing, 1*(1). Retrieved November 5, 2001, from http://aw.colostate.edu/connections/cw99/rosenberg.html

Ryle, G. (1949). *The concept of the mind.* London: Hutchinson.

Salvo, M. J. (1999). Trauma, narration, technology: User-ordered representation and the holocaust. *Computers and Composition, 16*(2), 283–301.

Schaafsma, D. (1993). *Eating on the street: Teaching literacy in a multicultural society.* Pittsburgh: University of Pittsburgh Press.

Schwalm, K. (1998). Statement. *Kairos: A Journal for Teachers of Writing in Webbed Environments, 3*(2). Retrieved November 5, 2001, from http://english.ttu.edu/kairos/3.2/coverweb/statements/schwalm.html

Selfe, C. L. (1989). Redefining literacy: The multilayered grammars of computers. In G. E. Hawisher

& C. L. Selfe (Eds.), *Critical perspectives in computers and composition instruction* (pp. 3–15). New York: Teachers College Press.

Selfe, C. L. (1992). Preparing teachers for the virtual age: The case for technology critics. In G. E. Hawisher & P. LeBlanc (Eds.), *Re-imagining composition in the virtual age* (pp. 24–42). Portmouth, NH: Boynton/Cook.

Selfe, C. L. (1996). Theorizing e-mail for the practice, instruction, and study of literacy. In P. Sullivan & J. Dautermann (Eds.), *Electronic literacies in the workplace: Technologies of writing* (pp. 255–293). Urbana, IL: National Council of Teachers of English.

Selfe, C. L. (1998, March). *Technology and literacy: A story about the perils of not paying attention.* Keynote address to the 1998 Convention of the Conference on College Composition and Communication, Chicago, IL.

Selfe, C. L. (1999a). Technology and literacy: A story about the perils of not paying attention. *College Composition and Communication, 50*(3), 411–436.

Selfe, C. L. (1999b). *Technology and literacy in the twenty-first century: The importance of paying attention.* Carbondale: Southern Illinois University Press.

Selfe, C. L. (1999c). Lest we think the revolution is a revolution: Images of technology and the nature of change. In G. E. Hawisher & C. L. Selfe (Eds.), *Passions, pedagogies, and 21st century technologies* (pp. 292–322). Logan: Utah State University Press.

Shauf, M. S. (2001). The problem of electronic argument: A humanist perspective. *Computers and Composition, 18*(1), 33–37.

Silverblatt, A. (1995). *Media literacy: Keys to interpreting media messages.* Westport, CT: Praeger.

Silverblatt, A. (1997). *Dictionary of media literacy.* Westport, CT: Greenwood.

Slatin, J. (1998). Statement. *Kairos: A Journal for Teachers of Writing in Webbed Environments, 3*(2). Retrieved November 5, 2001, from http://english.ttu.edu/kairos/3.2/coverweb/statements/slatin.html

Sloane, S. (1999). The haunting story of J. Genealogy as a critical category in understanding how a writer composes. In G. E. Hawisher & C. L. Selfe (Eds.), *Passions, pedagogies, and 21st century technologies* (pp. 49–65). Logan: Utah State University Press.

Snavely, L., & Cooper, N. (1997). The information literacy debate. *Journal of Academic Librarianship, 23,* 9–14.

Snyder, I. (Ed.). (1998). *Page to screen: Taking literacy into the electronic era.* London: Routledge.

Stoker, B. (1932). *Dracula.* New York: The Modern Library.

Stone, A. R. (1995). *The war of desire and technology at the close of the mechanical age.* Cambridge, MA: MIT Press.

Strasma, K. (1997). As you like it. *Kairos: A Journal for Teachers of Writing in Webbed Environments, 2*(1). Retrieved November 5, 2001, from http://english.ttu.edu/kairos/2.1/reviews/history/review.htm

Stuckey, J. E. (1991). *The violence of literacy.* Portsmouth, NH: Boynton/Cook.

Sullivan, P., & Dautermann, J. (Eds.). (1996). *Electronic literacies in the workplace: Technologies of writing.* Urbana, IL: National Council of Teachers of English.

Sullivan, P., & Porter, J. E. (1997). *Opening spaces: Writing technologies and critical research practices.* Norwood, NJ: Ablex.

Taylor, T. W., & Ward, I. (Eds.). (1998). *Literacy theory in the age of the Internet.* New York: Columbia University Press.

Thomas, L. G., & Knezek, D. G. (1998). *Technology literacy for the nation and its citizens.* Eugene, OR: International Society for Technology in Education.

Tuman, M. (1992). *Word perfect: Literacy in the computer age.* London: Falmer.

Ture, K., & Hamilton, C. V. (1967). *Black power: The politics of liberation.* New York: Vintage.

Turkle, S. (1995). *Life on the screen: Identity in the age of the Internet.* New York: Simon & Schuster.

Tyner, K. R. (1998). *Literacy in a digital world: Teaching and learning in the age of information.* Mahwah, NJ: Lawrence Erlbaum Associates.

Ulmer, G. (1998). Into electracy. In T. W. Taylor & I. Ward (Eds.), *Literacy theory in the age of the Internet* (pp. ix–xiii). New York: Columbia University Press.

United States Department of Education. (1996). *Getting America's students ready for the 21st century. Meeting the technology literacy challenge. A report to the nation on technology and education.* Washington, DC: Author.

Vitanza, V. (2001, March). *How electronic texts and journals will shape our professional work.* Plenary Address to the 2001 Convention of the Conference on College Composition and Communication, Denver, CO.

Walker, J., & Taylor, T.W. (1998). *The Columbia guide to online style.* New York: Columbia University Press.

Wambeam, C. A. (2000). *Participatory audience and hypertext: Emerging authors and their fan fiction.* Unpublished doctoral dissertation, New Mexico State University.

Warschauer, M. (1999). *Electronic literacies: Language, culture, and power in online education.* Mahwah, NJ: Lawrence Erlbaum Associates.

Welch, K. E. (1999). *Electric rhetoric: Classical rhetoric, oralism, and a new literacy.* Cambridge, MA: MIT Press.

Wysocki, A. (1998). Monitoring order: Visual desire, the organization of Web pages, and teaching the rules of design. *Kairos: A Journal for Teachers of Writing in Webbed Environments, 3*(2). Retrieved November 2, 2001, from http://english.ttu.edu/kairos/3.2

Wysocki, A., & Johnson-Eilola, J. (1999). Blinded by the letter: Why are we using literacy as a metaphor for everything else? In G. E. Hawisher & C. L. Selfe (Eds.), *Passions, pedagogies, and 21st century technologies* (pp. 349–368). Logan: Utah State University Press.

Author Index

Subject Index